Group and Ring Theoretic Properties
of Polycyclic Groups

T0214694

Algebra and Applications

Volume 10

Managing Editor:

Alain Verschoren
University of Antwerp, Belgium

Series Editors:

Alice Fialowski
Eötvös Loránd University, Hungary

Eric Friedlander
Northwestern University, USA

John Greenlees
Sheffield University, UK

Gerhard Hiss
Aachen University, Germany

Ieke Moerdijk
Utrecht University, The Netherlands

Idun Reiten
Norwegian University of Science and Technology, Norway

Christoph Schweigert
Hamburg University, Germany

Mina Teicher
Bar-Ilan University, Israel

Algebra and Applications aims to publish well written and carefully refereed monographs with up-to-date information about progress in all fields of algebra, its classical impact on commutative and noncommutative algebraic and differential geometry, K-theory and algebraic topology, as well as applications in related domains, such as number theory, homotopy and (co)homology theory, physics and discrete mathematics.

Particular emphasis will be put on state-of-the-art topics such as rings of differential operators, Lie algebras and super-algebras, group rings and algebras, C^*-algebras, Kac-Moody theory, arithmetic algebraic geometry, Hopf algebras and quantum groups, as well as their applications. In addition, Algebra and Applications will also publish monographs dedicated to computational aspects of these topics as well as algebraic and geometric methods in computer science.

B.A.F. Wehrfritz

Group and Ring Theoretic Properties of Polycyclic Groups

 Springer

B.A.F. Wehrfritz
School of Mathematical Sciences
Queen Mary, University of London
Mile End Road
London E1 4NS
UK
b.a.f.wehrfritz@qmw.ac.uk

ISBN 978-1-4471-2530-3   e-ISBN 978-1-84882-941-1
DOI 10.1007/978-1-84882-941-1
Springer London Dordrecht Heidelberg New York

British Library Cataloguing in Publication Data
A catalogue record for this book is available from the British Library

AMS codes: 20F16, 20F18, 20H20, 16P40, 13E05, 20C07, 16S34

Printed on acid-free paper

Springer is part of Springer Science+Business Media (www.springer.com)

Preface

A group is polycyclic if it can be built from a finite number of cyclic groups in a natural and simple way that we specify below. Polycyclic groups also crop up in a number of different contexts and can be defined in perhaps a surprising number of different ways. The most striking of these are, I think, that they are exactly the soluble groups with faithful matrix representations over the integers and are exactly the soluble groups satisfying the maximal condition on subgroups. As a result polycyclic groups appear in a large number of works that are not specifically about polycyclic groups, including works outside of group theory in areas such as ring theory, topology and even to a small extent geometry and number theory, the latter usually via arithmetic groups.

The theory of polycyclic groups started as a purely group theoretic exercise by Kurt Hirsch between the late 1930's to the early 1950's. I think he thought of it as a sort of analogue, a very superficial analogue, of Emmy Noether's theory of what are now called commutative Noetherian rings. It was Hirsch who first sparked my interest in polycyclic groups when I attended a year-long course he gave on polycyclic groups in 1964/5. This basic theory we cover in Chap. 2. The theory came of age with the introduction in the 1950's, by A.I. Mal'cev particularly, of the application of matrix group theory. This phase continued for some fifteen to twenty years. We describe how matrix theory is used to study polycyclic groups in Chaps. 4 and 5. Then came a halcyon period for the theory of polycyclic groups. In the 1970's and early 1980's two quite different and very powerful collections of techniques were developed by a number of people in a great rush of significant papers. The first can be briefly summarized as ring theoretic, especially non-commutative ring theoretic, with J.E. Roseblade being a major contributor. Explaining this is the prime object of my book and Chaps. 3, 6, 7, 8 and 9 are devoted to it. The theorems on polycyclic groups in these chapters seem to be of use as much outside the area of polycyclic groups as in it.

The second of these powerful collections of techniques can be loosely described as number, especially arithmetic, theoretic including the use of arithmetic groups. These were generally aimed more inside the theory of polycyclic groups rather than at those areas outside but involving at some level polycyclic groups. This is very well described and developed, along with very significant uses and theorems by Dan

Segal in the second half his book 'Polycyclic Groups'; Segal was a major player in this area. Thus I make no attempt to delve into this. Rather than rework Segal's splendid book, my aim is to complement it by concentrating on the ring theoretic area, an area that Segal does not try to cover. Since the second halves of our books have little if anything in common, this naturally affects our choice and order of development in the first halves. Also I have the opportunity to refer the reader to much later material even in the earlier part of the book.

Apart from advanced students of algebra, I want this book to be accessible to research workers in areas other than group theory who find themselves involved with polycyclic groups. I definitely intended to keep the book short and readable from start to finish. Towards the end of most sections and chapters I describe, often in some detail, the subsequent development of the topic, particularly with regard to more specialized results. Probably only a minority of readers will want to delve into the full proofs of any one of these discussions. My intention is that all readers should be able to read these discussions reasonably easily without getting too bogged down in the details and to get some idea of what sort of results are available in these areas, certainly enough of an idea for them to decide whether they need the full details, and then to inform them where to find these details.

The first half of this book, that is Chaps. 2 to 5, I covered in a one-semester University of London M.Sc. course. The second half I covered in a one-semester University of London course aimed at M.Phil. and Ph.D. students. Actually in both cases my book is a somewhat expanded version of these two courses with additional proofs and information. For the first half I assume that the reader has covered the equivalent of a one-semester course in group theory, but not necessarily recently. For readers whose group theory is inadequate or rusty I start the book with a Chap. 1, revising the general group theory that I need, with one or two less common results towards its end. I advise readers to start at Chap. 2 and only read Chap. 1 if they find the group theory really tough going. Other readers may well want to read up the odd result from Chap. 1, but I recommend they do so when they find it being quoted elsewhere in the book.

Particularly when it comes to the latter parts of the book I use quite a lot from ring and module theory. I give full references for theorems that I use. Unlike the group theory the ring theory that I quote tends to be just bits here and there so I have decided to leave it to the reader to read up whatever they find they need. I assume the reader has seen notions like ring, ideal, homomorphism, module, submodule and module homomorphism before. If not the reader would be well advised to read the first sections on rings and modules in some general algebra text such as the first volume of either P.M. Cohn's 'Algebra' or N. Jacobson's 'Basic Algebra' (or in one of the numerous other algebra texts). At various points I do need more advanced theorems. I hope I have stated what is being used sufficiently clearly that the reader can continue reading my proof with understanding without needing to refer immediately to another text. I stress that for a first reading it should not be necessary to actually have these standard texts to hand and in any case the reader may well have there own favorite texts containing the relevant material.

Queen Mary, University of London   B.A.F. Wehrfritz

Contents

Chapter 1
Some Basic Group Theory

We assume that the reader has some knowledge of basic group theory. Here we summarize what we use, mainly without proofs. Apart from reminding the reader of these ideas, it also enables us to introduce the reader to our notation. The exception to this is a couple of results at the end of this chapter that are unlikely to appear in first group theory courses. In the main we use them only once or perhaps twice in the latter half of the book, so readers might like to put off reading them until they actually need them. We present full proofs of these results.

We assume the reader knows the meaning of the following: group (usually written multiplicatively), subgroup ($H \leq G$ means H is a subgroup of the group G), normal subgroup ($H \lhd G$ means H is a normal subgroup of the group G), quotient (or factor) group, homomorphism, kernel, image, isomorphism, the isomorphism theorems, the correspondence theorem, the Jordan-Hölder Theorem, Sylow's Theorem, generators, abelian group, soluble group, nilpotent group, direct product of a finite number of groups, centralizer, normalizer ($C_G(X)$ denotes the centralizer of X in G and $N_G(X)$ the normalizer of X in G). All the above can be found, for example in the author's book (Wehrfritz 1999). Further we also use the following well-known result.

1.1 A finitely generated abelian group is a direct product (or sum if written additively) of a finite number of cyclic groups.

By definition a *free abelian* group is a direct sum of infinite cyclic groups.

Special Formulae

Let G be a group and x, y and z elements of G. Set $x^y = y^{-1}xy$ and

$$[x, y] = x^{-1}y^{-1}xy = x^{-1}x^y = y^{-x}y.$$

1.2

(a) $(xy)^z = x^z y^z$ and $x^{yz} = (x^y)^z$.

B.A.F. Wehrfritz, *Group and Ring Theoretic Properties of Polycyclic Groups*,
Algebra and Applications 10,
DOI 10.1007/978-1-84882-941-1_1, © Springer-Verlag London Limited 2009

(b) $[xy, z] = [x, z]^y [y, z]$ and $[x, yz] = [x, z][x, y]^z$.
(c) $[x, y]^{-1} = [y, x] = [x^y, y^{-1}]$.
(d) $[x, y^{-1}, z]^y \cdot [y, z^{-1}, x]^z \cdot [z, x^{-1}, y]^x = 1$.

If these are unfamilar, prove them by simply expanding both sides and comparing results. In the case of (d) note that by convention $[x, y, z]$ means $[[x, y], z]$; that is, always expand multiple commutators from left to right, unless extra brackets suggest otherwise. Note that direct expansion shows that

$$[x, y^{-1}, z]^y = u^{-1}v, \quad \text{where } u = xzx^{-1}yx \text{ and } v = yxy^{-1}zy.$$

Thus the next two commutators in (d) are $v^{-1}w$ and $w^{-1}u$ for $w = zyz^{-1}xz$, so clearly their product is 1.

If $H, K \leq G$, set $[H, K] = \langle [h, k] : h \in H \text{ and } k \in K \rangle$. Note that $[H, K] = [K, H]$ by 1.2(c). By definition $[H, K] = \langle 1 \rangle$ if and only if $H \leq C_G(K)$ (alternatively if and only if $K \leq C_G(H)$). If also $L \leq G$, then $[H, K, L]$ means $[[H, K], L]$.

1.3   If $H, K, L \triangleleft G$, then $[H, K] \triangleleft G$, $[H, K] \leq H \cap K$ and $[HK, L] = [H, L][K, L]$.

If $H, K \leq G$, then $[H, K] \triangleleft \langle H, K \rangle$.

1.4   If $H, K \leq G$, then $K \leq N_G(H)$ if and only if $[H, K] \leq H$.

1.5   **P. Hall's Three Subgroup Lemma** Let $X, Y, Z \leq G$ and $N \triangleleft G$. If $[X, Y, Z]$ and $[Y, Z, X]$ are contained in N then so is $[Z, X, Y]$.

To prove 1.5 use 1.2(d), but carefully.

1.6   *Direct Products*
Let G be a group and $H_\alpha \leq G$ for α in some index set I. The following are equivalent.

(a) Each $H_\alpha \triangleleft G$, $G = \langle H_\alpha : \alpha \in I \rangle$ and each $H_\alpha \cap \langle H_\beta : \beta \neq \alpha \rangle = \langle 1 \rangle$.
(b) $[H_\alpha, H_\beta] = \langle 1 \rangle$ for all $\alpha \neq \beta$ and for each $g \in G \backslash \langle 1 \rangle$ there exists a unique finite subset I_g of I and for each $\alpha \in I_g$ a unique element $h_\alpha \in H_\alpha \backslash \langle 1 \rangle$ such that $g = \prod h_\alpha$.

If the conditions in 1.6 hold we say G is the *direct product* (some say restricted direct product; if the groups are additive say direct sum) of the H_α and write $G = X_{\alpha \in I} H_\alpha$ (we also use notations like $G = H \times K$, $G = H_1 \times H_2 \times \cdots \times H_n$, $G = X_{1 \leq i \leq n} H_i$ etc.; for direct sum we replace the symbol $\times$ by $\oplus$ etc.).

1.7   *Split Extension $\equiv$ Semi-direct Product*
Suppose G is a group, $N \triangleleft G$ and $H \leq G$ with $G = HN$ ($= NH$) and $H \cap N = \langle 1 \rangle$. If $g \in G$, then $g = hn$ for some $h \in H$ and $n \in N$, uniquely, for if $g = h'n'$ where also $h' \in H$ and $n' \in N$, then $hn = h'n'$, $h^{-1}h' = n(n')^{-1} \in H \cap N = \langle 1 \rangle$ and $h = h'$ and $n = n'$. If $h \in H$ and $n \in N$, then $n^h \in N$ and

$$\gamma_h : n \mapsto n^h \quad \text{is an automorphism of } N \text{ (1.2a) and}$$

$$\gamma : h \mapsto \gamma_h \quad \text{is a homomorphism of } H \text{ into Aut } N.$$

(For $n\gamma_{hk} = n^{hk} = (n^h)^k = n\gamma_h\gamma_k$.) Also $h'n'.hn = (h'h)h^{-1}n'hn = (h'h)(n'\gamma_h.n)$.

Conversely let H and N be groups with $H \cap N = \langle 1 \rangle$ and with a homomorphism $\gamma : H \to \operatorname{Aut} N$. (If we are given such H and N, but with $H \cap N \neq \langle 1 \rangle$, then replace H by a copy of itself so that this intersection is $\langle 1 \rangle$; thus the condition $H \cap N = \langle 1 \rangle$ is a notational convenience rather than a real restriction.) Set $G = \{hn : h \in H, n \in N\}$, a set of formal products (i.e. ordered pairs (h, n)) hn. Define a multiplication on G by $h'n'.hn = (h'h)(n'(h\gamma).n)$. Then G becomes a group (check), $H = \{h1 : h \in H\}$ (using $1_N = 1_H = 1$) is a subgroup of G, $N = \{1n : n \in N\}$ is a normal subgroup of G and $G = HN$, $H \cap N = \langle 1 \rangle$ and $n(h\gamma) = n^h = n\gamma_h$ for $h \in H$ and $n \in N$.

The above two constructions, to within group isomorphism, are the converse of each other. We call G the split extension or semi-direct product of N and H. We write $G = H[N$ or $G = N]H$.

Soluble Groups

For any group G set $G' = [G, G]$ and inductively put $G^{(0)} = G$ and $G^{(n+1)} = (G^{(n)})'$ for $n = 0, 1, 2, \ldots$. Then $G \geq G' \geq G'' \geq \cdots \geq G^{(n)} \geq \cdots$ is the *derived series* of G and G' is the *derived subgroup* of G.

1.8 The group G is soluble of derived length $\leq d$, an integer, if any one of the following equivalent conditions hold.

(a) There exists a series $\langle 1 \rangle = N_0 \triangleleft N_1 \triangleleft N_2 \triangleleft \cdots \triangleleft N_d = G$ with each factor N_i / N_{i-1} abelian.
(b) There exists a series $\langle 1 \rangle = N_0 \triangleleft N_1 \triangleleft N_2 \triangleleft \cdots \triangleleft N_d = G$ with each $N_i \triangleleft G$ and each factor N_i / N_{i-1} abelian.
(c) $G^{(d)} = \langle 1 \rangle$.

Subgroups and images of soluble groups are soluble and extensions of soluble groups by soluble groups are soluble (meaning that if N is a normal subgroup of a group G with N and G/N soluble, then G is soluble). If M and N are soluble normal subgroups of some group G, then MN is also a soluble normal subgroup of G. An *abelian* group is just a soluble group of derived length at most 1. A *metabelian* group is a soluble group of derived length at most 2.

Nilpotent Groups

For any group G set $\gamma^1 G = G$, $\gamma^2 G = [\gamma^1 G, G] = G'$ and in general $\gamma^{i+1} G = [\gamma^i G, G]$. Each $\gamma^i G$ is normal in G and $\gamma^1 G \geq \gamma^2 G \geq \cdots \geq \gamma^i G \geq \cdots$ is the *lower central series* of G.

Denote the centre $C_G(G)$ of G by $\zeta_1(G)$ and inductively define $\zeta_i(G)$ by $\zeta_{i+1}(G)/\zeta_i(G) = \zeta_1(G/\zeta_i(G))$. Then $\langle 1 \rangle = \zeta_0(G) \leq \zeta_1(G) \leq \cdots \leq \zeta_i(G) \leq \cdots$ is the *upper central series* of G. (Both these series can be continued transfinitely by setting $\gamma^\lambda G = \bigcap_{\alpha < \lambda} \gamma^\alpha G$ and $\zeta_\lambda(G) = \bigcup_{\alpha < \lambda} \zeta_\alpha(G)$ for limit ordinals λ.)

1.9   The group G is nilpotent of class at most c, an integer, if any one of the following equivalent conditions hold.

(a) There exists a series $\langle 1 \rangle = N_0 \leq N_1 \leq N_2 \leq \cdots \leq N_c = G$ with $[N_i, G] \leq N_{i-1}$ for each i.
(b) There exists a series $\langle 1 \rangle = N_0 \leq N_1 \leq N_2 \leq \cdots \leq N_c = G$ with $N_{i-1} \triangleleft G$ and $N_i/N_{i-1} \leq \zeta_1(G/N_{i-1})$ for each $i > 0$.
(c) $\gamma^{c+1} G = \langle 1 \rangle$.
(d) $\zeta_c(G) = G$.

A series as in (a) or (b) of 1.9 is called a *central series*. The minimal such c is the *class* of G. Subgroups and images of nilpotent groups are nilpotent. However extensions of nilpotent groups by nilpotent groups need not be nilpotent (consider $G = \mathrm{Sym}(3)$ with $N = \mathrm{Alt}(3)$). For examples of nilpotent groups note that every finite group of prime power order is nilpotent, a very elementary result contained in virtually every undergraduate book on group theory, but see for example Wehrfritz (1999), 1.14.

1.10   (Hirsch 1937; Fitting 1938) If M and N are nilpotent normal subgroups of the group G, then MN is also nilpotent.

This is nothing like as trivial as the soluble analogue and some of the simplest proofs only work for finite groups. Thus we outline a proof here. Suppose $\gamma^{c+1} M = \langle 1 \rangle = \gamma^{d+1} N$. We prove that $\gamma^{c+d+1}(MN) = \langle 1 \rangle$. By 1.3

$$[AB, CD] = [A, C][A, D][B, C][B, D]$$

whenever A, B, C, and D are normal subgroups of G. Thus a simple induction yields

$$\gamma^{c+d+1}(MN) = \prod [A_0, A_1, \ldots, A_{c+d}]$$

where each A_i is either M or N and the product is over all possible choices of the A_i. Suppose $[A_0, A_1, \ldots, A_i] \leq \gamma^j M$. Using 1.3 we have $\gamma^j M \triangleleft G$, so

$$\text{if } \quad A_{i+1} = M, \quad \text{then} \quad [A_0, A_1, \ldots, A_{i+1}] \leq [\gamma^j M, M] \leq \gamma^{j+1} M$$

$$\text{and if } \quad A_{i+1} = N, \quad \text{then} \quad [A_0, A_1, \ldots, A_{i+1}] \leq [\gamma^j M, N] \leq \gamma^j M.$$

Therefore $[A_0, A_1, \ldots, A_{c+d}] \leq \gamma^r M \cap \gamma^s N$, where r is the number of A_i equal to M and s is the number equal to N. Clearly either $r > c$ or $s > d$. Thus each $[A_0, A_1, \ldots, A_{c+d}] = \langle 1 \rangle$ and the claim follows.

If G is any group the *Fitting subgroup* of G, denoted by $\eta_1(G)$ or $\mathrm{Fitt}(G)$, is the subgroup generated by all the nilpotent normal subgroups of G. If G is finite it follows from 1.10 that its Fitting subgroup is nilpotent. We will see later that the same applies to polycyclic groups. In these circumstances $\eta_1(G)$ is the unique maximal nilpotent normal subgroup of G. In general $\eta_1(G)$ need not be nilpotent (Exercise: construct such an example).

The Axiom of Choice

Let $(S, \leq)$ be a (non-empty) partially ordered set. A totally ordered subset of S is called a chain. If every chain C in S is bounded above, meaning that there exists b in S with $c \leq b$ for every c in C, we say S is inductively ordered. The following can be taken to be one of the axioms of set theory, as a version of the axiom of choice in fact. We can restrict to well-ordered ascending chains if we wish.

1.11   Zorn's Lemma Every non-empty inductively ordered set $(S, \leq)$ has a maximal member; that is there exists an m in S such that $m \leq s \in S$ implies $m = s$.

(Note that there is no need for m to be the maximum member of S or even to be unique.) Most of the time we shall not need Zorn's Lemma since we shall have a 'Noetherian condition', which delivers what we need, but there will be times when we are studying the influence of polycyclic groups on certain other structures (rings, modules, groups with a particular polycyclic image) when we will need to fall back on 1.11. We will be using this result in a very routine way and we give now a few examples of this method.

1.12   If A is an abelian subgroup of a group G there is a maximal abelian subgroup of G containing A.

For let S be the set of all abelian subgroups of G containing A, ordered by inclusion. Clearly S is non-empty and it is very easy to check that the union of a chain of abelian subgroups of G is an abelian subgroup of G. Thus S is inductively ordered. By 1.11 it has a maximal member and this maximal member is a maximal abelian subgroup of G containing A.

1.13   If P is a p-subgroup of a group G, there is a maximal p-subgroup of G containing P.

Here p is a prime. A p-group is a group whose elements all have orders powers of p. Then 1.13 follows from 1.11 by setting S equal to the set of all p-subgroups of G containing P, again ordered by inclusion.

1.14   Let G be a group with $g \in G \setminus \{1\}$. Then G contains a subgroup M maximal subject to $g \notin M$.

Here set S equal to the set of all subgroups of G not containing g. Check that S is inductively ordered with respect to inclusion. Note that we cannot prove this way that G has a maximal subgroup (meaning maximal-not-equal-to-G subgroup) and indeed it may not have such. For example the additive group of the rationals $\mathbf{Q}$ does not. We do, however have the following.

1.15   A finitely generated non-trivial group G has a maximal subgroup.

Let $G = \langle g_1, g_2, \ldots, g_n \rangle$, where n is minimal. Thus $g_1 \notin \langle g_2, \ldots, g_n \rangle < G$. We can use 1.11 to produce a subgroup M of G containing $\langle g_2, \ldots, g_n \rangle$ and maximal subject to not containing g_1. Any subgroup of G strictly containing M must contain g_1 and hence is G. Therefore M is a maximal subgroup of G.

The situation with rings is slightly different since we have two canonical elements to play with, 0 and 1. (All our rings will have an identity.)

1.16   Let $\mathbf{a} < R$ be an ideal of the ring R. Then R has a maximal ideal containing $\mathbf{a}$.

Let S be the set of all ideals $\neq R$ containing $\mathbf{a}$. Then subject to inclusion S is inductively ordered, since the union of a chain of ideals $\neq R$ does not contain 1 and therefore is an ideal $\neq R$ of R. The claim follows.

Exercise If A is an abelian normal subgroup of the group G, prove that A lies in a maximal abelian normal subgroup of G.

Exercise If P is a normal p-subgroup of G, prove that P lies in a maximal normal p-subgroup of G. Can you prove this without using Zorn's Lemma? (Your answer should be yes.)

Exercise If S is a soluble subgroup of a group G, does S always lie in a maximal soluble subgroup of G? (The answer is no; can you see why the Zorn's Lemma method above does not work.)

Exercise Prove that the additive group of the rationals and for p a prime the group of p-th power roots of unity in the complex numbers (that is, the *Prüfer p-group*) have no maximal subgroups. Can you derive one of these claims from the other?

Nilpotence

We complete this introductory chapter with some standard group theory connected with nilpotent groups. The Frattini subgroup $\Phi(G)$ of a group G is the intersection of all the maximal subgroups of G or G itself if none such exist.

1.17   (Frattini 1885a, 1885b) If G is a finite group, then $\Phi(G)$ is nilpotent.

Proof Let P be a Sylow p-subgroup of $N = \Phi(G)$. If $g \in G$, then P^g is also a Sylow p-subgroup of N and hence $P^{gx} = P$ for some $x \in N$ by Sylow's Theorem. Thus $gx \in N_G(P)$ and so $G = N.N_G(P)$. If $N_G(P) < G$, there exists a maximal subgroup M of G containing $N_G(P)$ and then $G = N.N_G(P) \leq \Phi(G).M \leq M < G$, a contradiction. Hence P is normal in G. It follows that N is the direct product of its Sylow subgroups and so is nilpotent. $\qquad\square$

1.18   (Schur 1904) If G is a group with $G/\zeta_1(G)$ finite, then G' is finite.

Proof Set $Z = \zeta_1(G)$ and let $t_1, t_2, \ldots, t_n$ be a transversal of Z to G, so $n = (G : Z)$. Now for all i and j there exist integers h&k and z_{ij}&z_i in Z with $t_i t_j = t_h z_{ij}$ and $t_i^{-1} = t_k z_i$. Let

$$Z_0 = \langle z_{ij}, z_i : 1 \leq i.j \leq n \rangle \leq Z.$$

Then $T = \bigcup_i t_i Z_0$ is a subgroup of G and $G/Z_0 = (T/Z_0) \times (Z/Z_0)$. Hence $G' \leq T$ and $G' \cap Z \leq T \cap Z = Z_0$. The latter is finitely generated note.

Let τ denote the transfer homomorphism of G into Z see, for example, Wehrfritz (1999), p. 71. Then $g\tau = g^n$ for $g \in Z$. Also $G' \leq \ker \tau$, so $(G' \cap Z)^n = \langle 1 \rangle$. Consequently $G' \cap Z$ and G' are finite. $\qquad \square$

Stability Groups

Let G be any group and $G = G_0 \supseteq G_1 \supseteq \cdots \supseteq G_r = \langle 1 \rangle$ a chain of subgroups of G. Suppose Γ is a group of automorphisms of G satisfying

$$(xG_i)^\gamma = xG_i \quad \text{for all } x \in G_{i-1}, \text{ all } i = 1, 2, \ldots, r \qquad (*)$$

and all $\gamma \in \Gamma$. We say that Γ stabilizes this chain of subgroups. (The set of all $\gamma \in \operatorname{Aut} G$ satisfying $(*)$ is a subgroup of $\operatorname{Aut} G$ called the stability group of the chain.)

For example, let N be a nilpotent normal subgroup of G and consider the chain

$$G = G_0 \supseteq G_1 \supseteq \cdots \supseteq G_c \supseteq G_{c+1} = \langle 1 \rangle$$

where $N = G_1, G_2, \ldots, G_{c+1}$ is a central series of N. Set $\Gamma = N/C_N(G)$. Then Γ stabilizes this chain (important special case: $N = G$ and $\Gamma = G/\zeta_1(G)$). For if $g \in N$ and $x \in G_{i-1}$, then $(xG_i)^g = x[x, g]G_i = xG_i$ for each $i \geq 1$.

In $(*)$ if we take $x = 1$ we see that $G_i^\gamma = G_i$, so Γ normalizes each G_i. If each G_i is normal in G_{i-1}, $(*)$ is equivalent to Γ inducing the trivial group on the factors G_{i-1}/G_i. The following notational trick is often useful. Regard G and Γ as subgroups of the split extension $\Gamma[G = \Gamma G$. Then x^γ becomes $\gamma^{-1}x\gamma$, $x^{-1}x^\gamma = [x, \gamma]$ and $(*)$ yields $[G_{i-1}, \Gamma] \leq G_i$ for each i. If also each G_i is normal in G then each G_i is normal in ΓG. We focus on this important special case, where each G_i is normal in G.

1.19 (Kaluzhnin 1950) Let $G = G_0 \supseteq G_1 \supseteq \cdots \supseteq G_r = \langle 1 \rangle$ be a series of normal subgroups of G. If $\Gamma \leq \operatorname{Aut} G$ stabilizes this series, then Γ is nilpotent of class less than r.

Proof We prove by induction on j that

$$[G_i, \Gamma_j] \leq G_{i+j} \quad \text{for all } i \text{ and } j, \text{ where } \Gamma_j = \gamma^j \Gamma. \qquad (**)$$

If $j = 1$, then $(**)$ becomes the definition of stability. Given $(**)$ for all i but some fixed $j \geq 1$, we have

$$[\Gamma, G_i, \Gamma_j] \leq [G_{i+1}, \Gamma_j] \leq G_{i+j+1} \quad \text{and}$$

$$[G_i, \Gamma_j, \Gamma] \leq [G_{i+j}, \Gamma] \leq G_{i+j+1} \quad \text{by } (**) \text{ for } j.$$

Now G_k is normal in $\Gamma[G$ for all k. Hence by P. Hall's Three Subgroup Lemma (1.5) we have $G_{i+j+1} \geq [\Gamma_j, \Gamma, G_i] = [\Gamma_{j+1}, G_i]$ as required. This proves (∗∗). In (∗∗) take $i = 0$ and $j = r$. Then $[G, \Gamma_r] = \langle 1 \rangle$, so $\Gamma_r = \langle 1 \rangle$ and Γ is nilpotent of class less then r. □

Hall (1957, 1958) has proved that even if the G_i are not necessarily normal in G, then Γ is nilpotent, but now its class can exceed $r - 1$ and is always at most $r(r - 1)/2$. As far as I know the exact bound is unknown. Hurley (1990) has improved Hall's bound. If $r = 4$ the correct bound is known to be 5 and not 6. Thus the above bound is not the best possible. We need to consider the case $r = 2$ in more detail.

1.20 Let N be a normal subgroup of the group G and let $\Gamma \leq \operatorname{Aut} G$ stabilize the chain $\langle 1 \rangle \leq N \leq G$. Let Z denote the centre of N.

(a) $[G, \Gamma] \leq Z$.
(b) If $g \in G$, then $\theta_g : \gamma \mapsto [g, \gamma] = g^{-1} g^\gamma$ is a homomorphism of Γ into Z.
(c) If $G = \langle X \rangle N$, e.g. if $X = G$, then $\theta = (\theta_x)_{x \in X} : \gamma \mapsto ([x, \gamma])_{x \in X}$ is an embedding of Γ into the cartesian product of $|X|$ copies of Z.

Proof (a) $[\Gamma, N, G] = [\langle 1 \rangle, G] = \langle 1 \rangle$ and $[N, G, \Gamma] \leq [N, \Gamma] = \langle 1 \rangle$. Hence by 1.5 again $[G, \Gamma, N] \leq \langle 1 \rangle$; that is, $[G, \Gamma]$ and N commute. But $[G, \Gamma] \leq N$. Therefore $[G, \Gamma] \leq Z$.

(b) $(\gamma \delta) \theta_g = [g, \gamma \delta] = [g, \delta][g, \gamma]^\delta = [g, \gamma][g, \delta] = (\gamma \theta_g)(\delta \theta_g)$, using (a), this being for all g in G and all γ and δ in Γ. Part (b) follows.

(c) Certainly θ is a homomorphism by (b). Let $\gamma \in \Gamma$ with $\gamma \theta = 1$. Then $[x, \gamma] = 1$ for all $x \in X$. Also $[N, \gamma] = \langle 1 \rangle$ and $\langle X, N \rangle = G$. Therefore $[G, \gamma] = \langle 1 \rangle$ and $\gamma = 1$. Thus Part (c) follows. □

1.21 Let $\Gamma \leq \operatorname{Aut} G$ stabilize the chain $\langle 1 \rangle = G_0 \leq G_1 \leq G_2 \leq \cdots \leq G_r = G$ of normal subgroups of the group G.

(a) If G_i / G_{i-1} is torsion-free for each $i < r$, then so too is Γ.
(b) If $G_i^q \leq G_{i-1}$ for each $i < r$ and some integer q, then Γ has exponent dividing q^{r-1}.
(c) If G is a finite p-group for some prime p, then Γ is also a finite p-group.

A group is *torsion-free* if each of its non-trivial elements has infinite order. The *exponent* of a group is the least common multiple of the orders of its elements (possibly infinite, even if these orders are all finite).

Exercise In 1.21 weaken the condition that each G_i is normal in G to each G_i is normal in G_{i-1}. Now prove (a), (c) and a version of (b) with a different bound.

Proof (a) & (b) We induct on r. Assume by induction that $\Gamma / C_\Gamma(G/G_1)$ is as required. Now $A = C_\Gamma(G/G_1)$ stabilizes $G \geq G_1 \geq \langle 1 \rangle$. Therefore A embeds into the cartesian product of copies of the centre of G_1 by 1.20. Hence A is torsion-free in the first case and satisfies $A^q = \langle 1 \rangle$ in the second. Parts (a) and (b) follow.

Let G be a finite p-group. Then $\Gamma \leq \operatorname{Aut} G$ is certainly finite. Also if $|G| = q$, a power of p, then each element of Γ has order dividing q^{r-1}, so Γ is a p-group. This proves (c).   □

Note that in Part (c) if G is just a p-group, then Γ need not be a p-group. For example, let $G = A \times B$ be the direct product of two copies A and B of the Prüfer p-group. Then the stability group of the chain $G \geq A \geq \langle 1 \rangle$ is torsion-free, being isomorphic the additive group of the p-adic integers $\mathbf{Z}_p$.

Exercise For any group G prove the following using 1.5 (P. Hall's Three Subgroup Lemma).

(a) $[\gamma^i G, \gamma^j G] \leq \gamma^{i+j} G$ for all $i, j \geq 1$.
(b) $[\gamma^i G, \zeta_j(G)] \leq \zeta_{j-i}(G)$ for $0 < i \leq j$.
(c) $\gamma^i(\gamma^j G) \leq \gamma^{ij} G$ for all $i, j \geq 1$.

1.22   Lemma (Neumann 1954) Suppose $G = x_1 H_1 \cup x_2 H_2 \cup \cdots \cup x_n H_n$, where the H_i are subgroups of the group G, the x_i are elements of the group G and n is finite. Then at least one of the H_i has finite index in G.

In fact, see Neumann (1954), at least one of the H_i has index at most n.

Proof We change notation slightly. Suppose that $H_1, H_2, \ldots, H_m$ are all distinct and $G = \bigcup_{i,j} x_{ij} H_i$, a union of finitely many cosets. If $m = 1$, clearly H_1 has finite index in G. If each coset of H_m is one of the $x_{mj} H_m$ clearly H_m has finite index in G. Suppose $x H_m$ is not one of the $x_{mj} H_m$. Since distinct cosets of H_m are disjoint we must have that $x H_m$ is contained in $\bigcup_{1 \leq i < m, j} x_{ij} H_i$. Hence

$$x_{mk} H_m \subseteq \bigcup_{1 \leq i < m, j} x_{mk} x^{-1} x_{ij} H_i$$

and so G is a union of finitely many cosets of $H_1, H_2, \ldots, H_{m-1}$. By induction on m at least one of the H_i for $i < m$ has finite index in G.   □

We conclude this preliminary chapter with the following more sophisticated result. We need it at only one point later in this book and readers may well prefer to leave its proof until that point. We pointed out earlier that an extension of a nilpotent group by a nilpotent group need not be nilpotent. The following is a weak but never-the-less useful substitute for this property.

1.23   Theorem (Hall 1957, 1958) Let G be a group and N a normal subgroup of G with N and G/N' nilpotent. Then G is nilpotent.

Proof We break the proof into three separate sections.

(a) Let A and B be G-modules (meaning A and B are additive abelian groups equipped with specified homomorphisms of G into $\text{Aut}\,A$ and $\text{Aut}\,B$ respectively). Let

$$\{0\} = A_0 \le A_1 \le \cdots \le A_m = A \quad \text{and} \quad \{0\} = B_0 \le B_1 \le \cdots \le B_n = B$$

be series of G-submodules of A and B that are stabilized by G. Then G stabilizes a series of length at most $m + n - 1$ in any G-image of $C = A \otimes_{\mathbf{Z}} B$.

Proof The action here of G on C is diagonal (that is $(a \otimes b)g = ag \otimes bg$, with the obvious notation). Set

$$C_k = \langle a \otimes b \colon \text{for all } a \in A_i \text{ and } b \in B_j \text{ such that } i + j = k \rangle$$
$$= \langle A_i \otimes B_j : i + j = k \rangle \le A \otimes B.$$

Then $\{0\} = C_1 \le C_2 \le \cdots \le C_{m+n} = C$. Also G stabilizes this series, for if $a \in A_i$ and $b \in B_j$ then

$$(a \otimes b)g = ag \otimes bg \quad \text{(so in particular this tensor lies in } A_i \otimes B_j \text{ and}$$
$$C_k \text{ is a } G\text{-submodule of } C),$$
$$= (a + a') \otimes (b + b') \quad \text{for some } a' \text{ in } A_{i-1} \text{ and } b' \text{ in } B_{j-1},$$
$$\in (a \otimes b) + C_{k-1}.$$

Further if $\phi : C \mapsto D$ is a G-homomorphism of C onto the G-module D, then G stabilizes the series $\{C_k\phi\}$ in D. $\qquad\square$

(b) Let N be a normal subgroup of G and set $A_i = \gamma^i N / \gamma^{i+1} N$. Then there is a surjection

$$N/N' \otimes_{\mathbf{Z}} A_{i-1} \to A_i$$

of G-modules for each $i \ge 1$.

Proof Certainly each A_i is a G-module via conjugation. If $g \in N$ and $x \in \gamma^{i-1}N$ then

$$[gN', x.\gamma^i N] \subseteq [g, x]\gamma^{i+1}N \quad \text{since } [N', \gamma^{i-1}N] \le \gamma^{i+1}N$$

by Part (a) of the exercise above. Thus

$$\phi : (gN', x.\gamma^i N) \mapsto [g, x]\gamma^{i+1}N \text{ of } A_1 \times A_{i-1} \to A_i$$

is well defined. Also ϕ is bilinear (over $\mathbf{Z}$) by 1.2, so we obtain $\phi^* : A_1 \otimes A_{i-1} \to A_i$. Finally ϕ^* is a G-map since $[x, y]^g = [x^g, y^g]$ and ϕ^* is onto since

$$\gamma^i N = [N, \gamma^{i-1}N] = \langle [g, x] : g \in N \text{ and } x \in \gamma^{i-1}N \rangle. \qquad\square$$

General Notation $[X, {}_1Y]$ means $[X, Y], [X, {}_2Y]$ means $[X, Y, Y]$ and in general for any positive integer n the expression $[X, {}_nY]$ means $[X, Y, Y, \ldots, Y]$, where Y appears n times.

(c) *The Proof of 1.23* Suppose G/N' is nilpotent of class c and N is nilpotent of class d, Then $[\gamma^1 N, {}_c G] \le N' = \gamma^2 N$. Suppose

$$[\gamma^{i-1} N, {}_{(i-1)c-(i-2)} G] \le \gamma^i N \quad \text{for some } i \ge 2.$$

By Parts (a) and (b)

$$[\gamma^i N, {}_{c+(i-1)c-(i-2)-1} G] \le \gamma^{i+1} N.$$

That is $[\gamma^i N, {}_{ic-i+1} G] \le \gamma^{i+1} N$. Thus this holds for all $i \ge 1$ by induction. But $\gamma^{d+1} N = \langle 1 \rangle$. Consequently $[N', {}_q G] = \langle 1 \rangle$ where

$$q = \sum_{2 \le i \le d} (ic - i + 1) = c(d(d+1)/2 - 1) - d(d-1)/2.$$

Also G/N' has class at most c. Thus $[G, {}_{c+q} G] = \langle 1 \rangle$ and G is nilpotent of class at most

$$c + q = cd(d+1)/2 - d(d-1)/2. \qquad \square$$

Remark Stewart (1966) showed that in fact G is nilpotent of class at most

$$cd + (c - 1)(d - 1).$$

Chapter 2
The Basic Theory of Polycyclic Groups

Group Classes

This is effectively just a language, developed by P. Hall in the 1950's and 60's to make certain types of group theoretic arguments more concise while highlighting the essential components of the proof.

A group class is a class $\mathbf{X}$ of groups such that $H \cong G \in \mathbf{X}$ implies $H \in \mathbf{X}$ (the main condition; effectively we are dealing with isomorphism classes of groups rather than the groups themselves) and such that $\langle 1 \rangle \in \mathbf{X}$ (a convenient convention). For certain commonly used classes we have special notations. These include the following.

$\mathbf{F}$	the class of all finite groups	$\mathbf{A}$	the class of all abelian groups
$\mathbf{G}$	the class of all finitely generated groups	$\mathbf{S}$	the class of all soluble groups
$\mathbf{G}_1$	the class of all cyclic ($=1$-generator) groups	$\mathbf{N}$	the class of all nilpotent groups
$\mathbf{I}$	the class of all trivial groups $\langle 1 \rangle$	$\mathbf{U}$	the class of all groups
$\mathbf{C}_n$	the class of cyclic groups of order n or 1		

An *operator* X on group classes is a function from group classes to group classes such that $X\mathbf{X} \subseteq X\mathbf{Y}$ whenever $\mathbf{X} \subseteq \mathbf{Y}$, such that $\mathbf{X} \subseteq X\mathbf{X}$ and such that $X\mathbf{I} = \mathbf{I}$ (necessarily $X\mathbf{U} = \mathbf{U}$). Operators multiply in the obvious way; viz. $YX\mathbf{X} = Y(X\mathbf{X})$. This multiplication of operators is associative but not commutative. If $X\mathbf{X} = \mathbf{X}$ we say $\mathbf{X}$ is X-closed. We say X is a *closure operator* if $X = X^2$; that is if $X(X\mathbf{X}) = X\mathbf{X}$, in other words if $X\mathbf{X}$ is X-closed for all $\mathbf{X}$.

We have the following commonly used closure operators. S is the subgroup operator; $S\mathbf{X}$ is the class of all subgroups of $\mathbf{X}$-groups. Now subgroups of soluble groups are soluble. Thus $SS = S$ and so S is S-closed (or say S is subgroup closed). Q is

B.A.F. Wehrfritz, *Group and Ring Theoretic Properties of Polycyclic Groups*, Algebra and Applications 10, DOI 10.1007/978-1-84882-941-1_2, © Springer-Verlag London Limited 2009

the quotient operator (some authors write H for Q); $Q\mathbf{X}$ is the class of all (homomorphic) images of $\mathbf{X}$-groups. Images of soluble groups are soluble. Hence $Q\mathbf{S} = \mathbf{S}$ and $\mathbf{S}$ is Q-closed (or image closed). In the same way we have that $\mathbf{A}, \mathbf{N}, \mathbf{F}, \mathbf{G}_1, \mathbf{U}$ and $\mathbf{I}$ are S-closed and Q-closed; $\mathbf{G}$ is Q-closed but not S-closed (try a free group).

Exercise Find a soluble example; we will see such examples later.

There is a product of group classes. If $\mathbf{X}$ and $\mathbf{Y}$ are group classes, then $\mathbf{XY}$ is the class of all groups G with a normal subgroup N such that N is an $\mathbf{X}$-group and G/N is a $\mathbf{Y}$-group. We say G in $\mathbf{XY}$ is an extension of an $\mathbf{X}$-group by a $\mathbf{Y}$-group or just say G is $\mathbf{X}$ by $\mathbf{Y}$. For example $\mathbf{SS} = \mathbf{S}$ while $\mathbf{S} \supset \mathbf{NN} \supset \mathbf{N}$. Clearly $\mathbf{IX} = \mathbf{XI} = \mathbf{X}$ and $\mathbf{UX} = \mathbf{XU} = \mathbf{U}$. Warning: in general $(\mathbf{XY})\mathbf{Z} \neq \mathbf{X}(\mathbf{YZ})$. For example, the alternating group Alt(4) of order 12 lies in $(\mathbf{G}_1\mathbf{G}_1)\mathbf{G}_1$ but not in $\mathbf{G}_1(\mathbf{G}_1\mathbf{G}_1)$.

Exercise For any group classes $\mathbf{X}$, $\mathbf{Y}$ and $\mathbf{Z}$, prove that $(\mathbf{XY})\mathbf{Z}$ always contains $\mathbf{X}(\mathbf{YZ})$.

An important closure operator, especially for us here, is the poly operator P (some authors write E for P). If G is a group with a series of subgroups

$$\langle 1 \rangle = G_0 \lhd G_1 \lhd \cdots \lhd G_r = G, \qquad\qquad (*)$$

where r is an integer, and if each $G_{i+1}/G_i \in \mathbf{X}$, we write $G \in P\mathbf{X}$ and say G is poly $\mathbf{X}$. Then P is a closure operator; further $\mathbf{X}$ is P-closed if and only if whenever $N \lhd G \in \mathbf{U}$ with N and G/N in $\mathbf{X}$, then $G \in \mathbf{X}$. For example, $P\mathbf{A} = \mathbf{S}$, $P\mathbf{N} = \mathbf{S}$ and $P\mathbf{S} = \mathbf{S}$. The primary object of study in this book is $\mathbf{P} = P\mathbf{G}_1$, the class of polycyclic groups.

2.1 For any group class $\mathbf{X}$ we have the following.

(a) $SP\mathbf{X} \subseteq PS\mathbf{X}$ (we write $SP \leq PS$).
(b) $QP\mathbf{X} \subseteq PQ\mathbf{X}$ (we write $QP \leq PQ$).

Proof (a) If H is a subgroup of G and we are given the series $(*)$ above, then

$$\langle 1 \rangle = G_0 \cap H \lhd G_1 \cap H \lhd \cdots \lhd G_r \cap H = H$$

is a series for H and

$$(G_i \cap H)/(G_{i-1} \cap H) = (G_i \cap H)/(G_{i-1} \cap (G_i \cap H))$$

$$\cong (G_i \cap H)G_{i-1}/G_{i-1} \leq G_i/G_{i-1}.$$

If follows that if $G \in P\mathbf{X}$ then $H \in PS\mathbf{X}$ and hence that $SP \leq PS$.

(b) If N is a normal subgroup of G and again we are given $(*)$, then

$$\langle 1 \rangle = G_0 N/N \lhd G_1 N/N \lhd \cdots \lhd G_r N/N = G/N$$

is a series for G/N and

$$(G_i N/N)/(G_{i-1} N/N) \cong G_i N/G_{i-1} N \cong G_i/(G_i \cap G_{i-1} N)$$
$$= G_{i//}G_{i-1}(G_i \cap N),$$

which is an image of G_i/G_{i-1}. Thus if $G \in P\mathbf{X}$ then $G/N \in PQ\mathbf{X}$ and so $QP \leq PQ$. $\qquad\square$

2.2   Corollary Subgroups and images of polycyclic groups are polycyclic.

For $SP = SP\mathbf{G}_1 \subseteq PS\mathbf{G}_1 = P\mathbf{G}_1 = \mathbf{P}$. Similarly $QP = \mathbf{P}$.

If $\mathbf{X}$ is any group class, a group G is *residually* $\mathbf{X}$ (and we write $G \in R\mathbf{X}$) if for each g in $G\backslash\langle 1\rangle$ there is a normal subgroup N of G such that $g \notin N$ and $G/N \in \mathbf{X}$; that is, if the intersection of the normal subgroups N of G with G/N an $\mathbf{X}$-group is trivial. Equivalently, if for each $g \in G\backslash\langle 1\rangle$ there is a homomorphism ϕ of G onto an $\mathbf{X}$ group with $g\phi \neq 1$. It is easy to check that R is another closure operator.

Exercise If $\mathbf{X}$ is an S-closed class prove that the group $G \in R\mathbf{X}$ if and only if G is isomorphic to a subgroup of a cartesian product of $\mathbf{X}$-groups.

There is one further standard operator that we shall have occasional recourse to, namely the *local* operator. A group $G \in L\mathbf{X}$, that is lies in the class of *locally* $\mathbf{X}$-groups, if for every finite *subset* F of G there is an $\mathbf{X}$-subgroup H of G containing F. Again L is a closure operator. Note that this is not quite the same as demanding that the finitely generated subgroups of G be $\mathbf{X}$-groups. If $\mathbf{X}$ is S-closed then $G \in L\mathbf{X}$ if and only if every finitely generated subgroup of G is an X-group (simply replace H by the subgroup generated by F). Thus a group is locally soluble (resp. locally nilpotent; resp. locally finite) if each of its finitely generated subgroups is soluble (resp. nilpotent; resp. finite). Clearly $L\mathbf{A} = \mathbf{A}$.

Exercise Determine the classes $L\mathbf{G}_1$ (consider the sections of the additive group of the rationals) and $L\mathbf{G}$.

Polycyclic Groups

We are now ready to start on our main area of study. Just to recall what we have above, a group G is polycyclic if it has a series of finite length with cyclic factors and subgroups and images of polycyclic groups are polycyclic. Further a polycyclic-by-finite group is a group with a polycyclic normal subgroup of finite index. We will see below that for many purposes the class $\mathbf{PF}$ of polycyclic-by-finite groups is a more natural object of study than $\mathbf{P}$ itself. It follows from the polycyclic case that subgroups and images of polycyclic-by-finite groups are also polycyclic-by-finite.

2.3   Let G be a group. The following are equivalent.

(a) Every subgroup of G is finitely generated.

(b) Every ascending chain $G_1 \leq G_2 \leq \cdots \leq G_i \leq \cdots$, where $i = 1, 2, \ldots,$ of subgroups of G contains only finitely many distinct members.
(c) Every non-empty set of subgroups of G has a maximal member.

Proof This is a special case of a well-known bit of universal algebra depending ultimately on the axiom of choice.

(a) implies (b). The union H of the G_i is finitely generated by (a), say by $g_1, g_2, \ldots, g_n$. Then there exists j such that G_j contains all the g_k. Thus $G_i = G_j$ for $i \geq j$.

(b) implies (c). Let S be a non-empty set of subgroups of G with no maximal member. There exists G_1 in S and G_1 is not a maximal member of S. Hence there exists G_2 in S with $G_1 < G_2$. Again G_2 is not maximal so there exists G_3 in S with $G_2 < G_3$. Keep going infinitely often (this is where the axiom of choice kicks in). We obtain a contradiction of (b).

(c) implies (a). Consider $H \leq G$. Let S denote the set of all finitely generated subgroups of H. Then $\langle 1 \rangle$ lies in S, so S is not empty. By (c) there is a maximal member M of S. If $h \in H$, then $\langle M, h \rangle$ lies in S and contains M. Consequently $M = \langle M, h \rangle$ and $h \in M$. Therefore $H = M$, which lies in S, so H is finitely generated. $\square$

If the group G satisfies the conditions of 2.3 we say that G satisfies the *maximal condition on subgroups*, a phrase which we shorten to max, and write $G \in \mathbf{Max}$; that is $\mathbf{Max}$ denotes the class of groups satisfying max. Clearly $\mathbf{G_1} \cup \mathbf{F} \subseteq \mathbf{Max}$.

2.4

(a) $\mathbf{G}$ is Q- and P-closed.
(b) $\mathbf{Max}$ is S-, Q- and P-closed.

The class $\mathbf{G}$ is not S-closed: let G denote the following subgroup of 2 by 2 rational matrices.

$$\left\langle \begin{pmatrix} 1 & 0 \\ 1 & 1 \end{pmatrix}, \begin{pmatrix} 2 & 0 \\ 0 & 1 \end{pmatrix} \right\rangle.$$

Then G contains the subgroup

$$\left\langle \begin{pmatrix} 1 & 0 \\ r2^s & 1 \end{pmatrix} : \text{all integers } r, s \right\rangle,$$

which is isomorphic to the non-finitely generated, additive group of the ring $\mathbf{Z}[1/2]$.

Proof of 2.4 (a) This is obvious.

(b) The S-closure follows from 2.3(a) and the Q-closure from 2.3(a) and part (a). Using 2.1 and 2.3 we have

$$SP\mathbf{Max} \subseteq PS\mathbf{Max} \subseteq P\mathbf{G} = \mathbf{G} \quad \text{and} \quad P\mathbf{Max} \subseteq \mathbf{Max}. \qquad \square$$

2.5   Corollary (Hirsch 1938a, 1938b) $P(\mathbf{G_1} \cup \mathbf{F}) \subseteq \mathbf{Max}$ and $\mathbf{G} \cap \mathbf{A} \subseteq P\mathbf{G_1} \subseteq \mathbf{Max}$.

Exercise Prove that $\mathbf{P} = \mathbf{S} \cap \mathbf{Max} \subseteq \mathbf{S} \cap \mathbf{G}$ and $\mathbf{G} \cap \mathbf{A} = \mathbf{Max} \cap \mathbf{A}$.

For a long time the $P(\mathbf{G}_1 \cup \mathbf{F})$ groups were the only known groups with max. Then Ol'shanskii (1979) produced examples of infinite groups all of whose proper subgroups are cyclic of prime order. Such groups clearly satisfy max (and also min for that matter). As time went by more examples of this general type were discovered. For example groups were constructed with the proper subgroups all of the same prime order. None of these constructions are easy. See Ol'shanskii (1991) for an account of this.

2.6   Theorem (Hirsch 1938a, 1938b, 1946) Let G be a group. The following are equivalent.

(a) G has a series $\langle 1 \rangle = G_0 \lhd G_1 \lhd \cdots \lhd G_r = G$ with each factor cyclic or finite.
(b) G has a series $\langle 1 \rangle = H_0 \lhd H_1 \lhd \cdots \lhd H_s \lhd G$ with each H_i / H_{i-1} infinite cyclic and G/H_s finite.
(c) G has a series $\langle 1 \rangle = K_0 \lhd K_1 \lhd \cdots \lhd K_t \lhd G$ with each $K_i \lhd G$, with each K_i / K_{i-1} free abelian of finite rank and with G/K_t finite.

2.7   Corollary The following hold.

(a) $P(\mathbf{G}_1 \cup \mathbf{F}) = \mathbf{PF}$. (Thus poly (cyclic or finite) groups are polycyclic by finite.)
(b) Polycyclic groups are torsion-free by finite.
(c) If G is a polycyclic-by-finite group, the finite subgroups of G have bounded order.
(d) $\mathbf{P} \subseteq (\mathbf{PZ})\mathbf{F}$; that is, polycyclic groups are (poly infinite-cyclic) by finite.

Proof of 2.6 Trivially (c) implies (b) and (b) implies (a). Suppose (a) holds. We prove (b), which is the main part of the proof. We induct on r. By induction applied to G_{r-1} we may assume that G_i/G_{i-1} is infinite cyclic for $i \leq r - 2$ and G_{r-1}/G_{r-2} is finite, of order n say. Set $N = G_{r-1}^n = \langle g^n : g \in G_{r-1} \rangle$. Then N is normal in G and is contained in G_{r-2} with G_{r-1}/N finite. Also

$$\langle 1 \rangle \lhd N \cap G_1 \lhd N \cap G_2 \lhd \cdots \lhd N \cap G_{r-3} \lhd N$$

is a series for N with infinite cyclic factors. If G/G_{r-1} is finite, then G/N is finite and (b) holds.

Suppose G/G_{r-1} is infinite (and hence cyclic). Set $P = G_{r-1}/N$ and $Q = G/N$. Put $C = C_Q(P)$. Then P is finite and Q/C embeds into Aut P, so Q/C is finite. Also $C/(C \cap P)$ is infinite cyclic, being isomorphic to a non-trivial subgroup of $Q/P \cong G/G_{r-1}$, and $C \cap P$ is central in C. Thus C is abelian and if $m = |C \cap P|$ then C^m is infinite cyclic and normal in Q and $(Q : C^m) = (Q : C)(C : C^m)$ is finite. Set $L/N = C^m$. Then L/N is infinite cyclic, G/L is finite and

$$\langle 1 \rangle \lhd N \cap G_1 \lhd N \cap G_2 \lhd \cdots \lhd N \cap G_{r-3} \lhd N \lhd L \lhd G$$

is a series of the type required by (b).

It remains to prove that (b) implies (c). Now in (b) the group H_s is clearly soluble; let K_1 be the last but one term of its derived series. Then K_1 is polycyclic, abelian, torsion-free and normal in G. In particular K_1 is free abelian of finite rank. Repeat with G/K_1 to define K_2/K_1. Continue in this way. By 2.5 this process stops, say at K_t, after a finite number of steps. Clearly then G/K_t is finite.   □

The Hirsch Number

This numerical invariant is an important tool for inductive proofs involving poly-cyclic groups. Let $\langle 1 \rangle = G_0 \lhd G_1 \lhd \cdots \lhd G_r = G$ be a series of the group G with factors cyclic or finite. Now if A is a subgroup of the infinite cyclic group B, then either $A = \langle 1 \rangle$ and B/A is infinite cyclic, or A is infinite cyclic and B/A is finite. Thus any refinement of the above series has the same number of infinite cyclic factors as the original series. Now any two series have isomorphic refinements (a theorem of Schreier from 1928, e.g. see Wehrfritz (1999), Theorem 1.3). Thus any two poly (cyclic or finite) series of a group G have the same number of infinite cyclic factors (Hirsch 1938a, 1938b). This invariant of the polycyclic-by-finite group G is called the Hirsch number of G; we denote it by $h(G)$. Frequently arguments use induction on this non-negative integer $h(G)$.

Exercise If H is a normal subgroup and K is a subgroup of the polycyclic-by-finite group G, prove that

$$h(G) = h(H) + h(G/H) \quad \text{and} \quad h(K) \leq h(G).$$

Polycyclic groups are greatly influenced by their finite images. Our first example of this is the following further theorem of Hirsch.

2.8 (Hirsch 1946) If every finite image of the polycyclic-by-finite group G is nilpotent, then G is nilpotent.

Proof Clearly we may assume that G is infinite, so by 2.6 there is a non-trivial torsion-free abelian normal subgroup A of G. By induction on $h(G)$ each G/A^p is nilpotent for each prime p. If $r = h(A)$ (equivalently $r = \text{rank } A$), then $(A : A^p) = p^r$ and $[A, {}_r G] \leq \bigcap_p A^p = \langle 1 \rangle$. Also G/A is nilpotent. Therefore G is nilpotent. □

Developments of 2.8

This theorem of Hirsch stimulated a long chain of results. Baer (1957) proved that a polycyclic-by-finite group with all its finite images supersoluble is itself super-soluble. Unlike 2.8 this is not an elementary result; it involves some substantial number theory. For a proof see Segal (1983) p. 54, Theorem 1 or Wehrfritz (1973a) Theorem 11.11. Independently Robinson (1970) and Wehrfritz (1970) proved that a finitely generated soluble group with each of its finite images nilpotent is nilpotent (see Robinson (1980) 15.5.3 for a proof). Platonov (1966) and Wehrfritz (1968), again independently and with different approaches, proved that a finitely generated group of matrices over a field with each of its finite images nilpotent, is nilpotent, see Wehrfritz (1973a) 4.16 and 10.5 for the two different approaches. The example given between 2.4 and 2.5 above is a finitely generated soluble (even metabelian) matrix group with all its finite images supersoluble that is not supersoluble. The more complex situation here is completely analysed in Segal (1975a). More recently Endimioni (1998) proved that if every finite image of a polycyclic group G has a series of length n with nilpotent factors, then so does G. For $n = 1$ this becomes Hirsch's theorem 2.8.

2.9   Theorem (Mal'cev 1958) Let H be any subgroup of the polycyclic-by-finite group G. Then H is the intersection of all the subgroups of G of finite index containing H.

Proof Note first that if B is a finitely generated abelian group, then $\bigcap_{r \geq 1} B^r = \langle 1 \rangle$, for if B is the direct product $\times_i \langle b_i \rangle$, then $B^r = \times_i \langle b_i^r \rangle$ and $\bigcap_r B^r = \times_i \bigcap_r \langle b_i^r \rangle = \langle 1 \rangle$.

   Again we induct on $h(G)$. If G is finite there is nothing to prove, so assume otherwise. Then we have a non-trivial torsion-free abelian normal subgroup A of G. Set $H^\wedge$ equal to the intersection of all the subgroups of G of finite index containing H. By induction applied to each G/A^r we have

$$H \leq H^\wedge \leq \bigcap_{r \geq 1} H A^r \leq H A.$$

Thus

$$H^\wedge \leq H\left(A \cap \bigcap_{r \geq 1} H A^r \right) = H\left(\bigcap_{r \geq 1} (H \cap A) A^r \right) = H(H \cap A) = H,$$

since $A/(H \cap A)$ is a finitely generated abelian group. Therefore $H = H^\wedge$.   □

2.10   Corollary (Hirsch 1952) If G is a polycyclic-by-finite group, then G is residually finite.

Proof If $L \leq G$, then $L_G = \bigcap_{g \in G} L^g$ is the largest normal subgroup of G contained in L. If L has finite index in G, then $G = \bigcup_{x \in X} Lx$ for some finite subset X of G, $L_G = \bigcap_{x \in X} L^x$ and L_G has finite index in G. Therefore

$$H^\wedge = \bigcap (L : H \leq L \leq G \text{ with } (G : L) \text{ finite})$$

$$= \bigcap (HN : N \triangleleft G \text{ with } (G : N) \text{ finite}).$$

In particular if G is polycyclic-by-finite and if $H = \langle 1 \rangle$, this and 2.9 say that G is residually finite. $\square$

There are too many results related to 2.10 to be able to mention more than just a couple of them. By a very deep theorem (Roseblade 1973, 1976 with Hall 1959 and Jategaonkar 1974) a finitely generated abelian-by-polycyclic-by-finite group is always residually finite. Much of the latter part of this book revolves around this result, see 9.13 below. Also (Mal'cev 1940) any finitely generated matrix group (over a field) is residually finite (see Wehrfritz 1973a 4.2 for a proof).

If p is a prime and π a set of primes, we denote the class of finite p-groups by $\mathbf{F}_{\{p\}}$ and the class of finite π-groups by $\mathbf{F}_\pi$

2.11   Theorem (Shmel'kin 1968 and Wehrfritz 1970 independently)

$$\mathbf{PF} \subseteq \bigcap_{\text{primes } p} ((R\mathbf{F}_{\{p\}})\mathbf{F}).$$

That is, if G is a polycyclic-by-finite group and p is any prime, then there exists N_p normal of finite index in G such that N_p is residually a finite p-group.

We will give a very easy proof of this theorem in Chap. 4, see 4.10.

Exercise Give a direct proof of 2.11 from the above—see Shmel'kin (1968).

2.12   Corollary (Learner 1964) If G is polycyclic-by-finite there exists a finite set π of primes such that G is residually a finite π-group (that is, $G \in R\mathbf{F}_\pi$).

Proof With N_2 as in the statement of 2.11, let π denote the set of prime divisors of $2(G : N_2)$. $\square$

2.13   Theorem (Hirsch 1938a, 1938b) $\mathbf{G} \cap \mathbf{N} \subseteq \mathbf{P}$. That is, finitely generated nilpotent groups are polycyclic.

Proof Let $\gamma^c G$ be the last non-trivial term of the lower central series of G. By induction we may assume that $G/\gamma^c G$ is polycyclic. In particular $\gamma^{c-1}G/\gamma^c G$ is finitely generated. Suppose $G = \langle x_1, x_2, \ldots, x_m \rangle$ and $\gamma^{c-1}G = \langle y_1, y_2, \ldots, y_n \rangle \gamma^c G$. Then

$$\gamma^c G = [\langle x_1, x_2, \ldots, x_m \rangle, \langle y_1, y_2, \ldots, y_n \rangle] = \langle [x_i, y_j] : 1 \le i \le m \ \& \ 1 \le j \le n \rangle$$

by 1.2(b), since $\gamma^c G$ is central in G. Thus $\gamma^c G$ is finitely generated and it follows that G is polycyclic.

Alternatively if G is any group with G/G' finitely generated then part (b) of the proof of 1.23 and a trivial induction shows that $\gamma^n G/\gamma^{n+1}G$ is finitely generated for each $n \ge 1$ and then 2.13 follows easily. (A third proof will be given in Chap. 3.) $\square$

2.14

(a) If G is a nilpotent group, then G has normalizer condition, meaning that $H < G$ implies that $H < N_G(H)$.

(b) If G is a group with normalizer condition, then the set of all elements of G of finite order is a subgroup of G and is a direct product of p-groups, one for each prime p.

(c) (Hirsch 1938a, 1938b) Finitely generated nilpotent groups are finite by torsion-free.

Proof (a) The group G has a central series $\langle 1 \rangle = G_0 \leq G_1 \leq \cdots \leq G_r = G$ of finite length. Since $H \neq G$ there exists s with $G_s \leq H$ and G_{s+1} not contained in H. Also $[H, G_{s+1}] \leq [G, G_{s+1}] \leq G_s \leq H$, so $G_{s+1} \leq N_G(H)$. Therefore $H \neq N_G(H)$.

(b) Let p be a prime and let G_p be any maximal p-subgroup of G (such exists by 1.13). Then G_p is characteristic in $N_G(G_p)$, so G_p is normal in $N_G(N_G(G_p))$. The latter is therefore equal to $N_G(G_p)$ and hence is G by hypothesis. Thus G_p is normal in G and hence is the unique maximal p-subgroup of G. It follows that the set of elements of G if finite order is equal to $\langle G_p : p \text{ prime} \rangle$ and that the latter is just the direct product of the G_p.

(c) This follows from 2.13, 2.7(b) and parts (a) and (b). $\square$

Note that polycyclic groups need not be finite by torsion-free. For example the infinite dihedral group

$$D_\infty = \langle a, x : a^x = a^{-1}, x^2 = 1 \rangle = \langle x \rangle [\langle a \rangle$$

is clearly polycyclic, being infinite-cyclic by cyclic-of-order-2. Also

$$(xa)^2 = xaxa = x^{-1}axa = a^{-1}a = 1.$$

Thus xa and x both have finite order (namely 2) and clearly $D_\infty = \langle xa, x \rangle$, so D_∞ is not finite by torsion-free.

Exercise Prove that D_∞ is isomorphic to the following matrix group over the integers

$$\left\langle \begin{pmatrix} 1 & 0 \\ 1 & 1 \end{pmatrix}, \begin{pmatrix} -1 & 0 \\ 0 & 1 \end{pmatrix} \right\rangle.$$

2.15   Let G be a nilpotent group. If $Z = \zeta_1(G)$ is torsion-free, then so are G, G/Z and each $\zeta_i(G)/\zeta_{i-1}(G)$. If $Z^n = \langle 1 \rangle$, then $\zeta_i(G)^n \leq \zeta_{i-1}(G)$ for all $i \geq 1$ and G has finite exponent (meaning that $G^m = \langle 1 \rangle$ for some positive integer m).

Proof Suppose Z is torsion-free and let $z \in \zeta_2(G)$ with $z^r \in Z$. If $g \in G$, then $1 = [z^r, g] = [z, g]^r$ by 1.2(b), since $[z, g] \in Z$ is central. Then $[z, g] = 1$, for all g, so $z \in Z$ and $\zeta_2(G)/\zeta_1(G)$ is torsion-free. By an elementary induction each $\zeta_i(G)/\zeta_{i-1}(G)$ is torsion-free. Consequently so too are G and G/Z.

Exercise Poly torsion-free is torsion-free.

Now assume that $Z^n = \langle 1 \rangle$. If $z \in \zeta_2(G)$, then $[z^n, g] = [z, g]^n = 1$. It follows that $\zeta_2(G)^n \leq \zeta_1(G)$. The remaining claims follow easily by induction. $\square$

If G is any group we denote by $\tau(G)$ the unique maximal periodic normal subgroup of G. (A *periodic* group is one all of whose elements have finite order.)

Exercise Any group has a unique maximal periodic normal subgroup.

Warning: in general $\tau(G)$ is not equal to the set of elements in G of finite order and $G/\tau(G)$ need not be torsion-free. For example $\tau(D_\infty) = \langle 1 \rangle$.

2.16   Theorem (Gruenberg 1957) Let G be a finitely generated nilpotent group.

(a) If $\tau(G) = \langle 1 \rangle$ (equivalently by 2.14, if G is torsion-free), then $G \in \bigcap_p RF_{\{p\}}$, the intersection being over all primes p; that is, G is residually a finite p-group for every prime p.
(b) If $\tau(G) \neq \langle 1 \rangle$ is a π-group, then $G \in RF_\pi$.

Proof (a) Set $Z = \zeta_1(G)$. By 2.15 and induction on the nilpotency class $G/Z \in RF_{\{p\}}$. Let $z \in Z \backslash \langle 1 \rangle$. Now Z is free abelian of finite rank by 2.13, so $z \notin Z^q$ for $q = p^i$ and some i. Pick (by 2.5) N with $Z^q \leq N \triangleleft G$ and maximal in G subject to $z \notin N$. Now G/N is polycyclic (2.13), so G/N is residually finite by 2.10. Hence there is $M \geq N$, normal of finite index in G with $zN \notin M/N$, that is with $z \notin M$. By the choice of N we have $M = N$, so G/N is finite. By 2.14 the group G/N is a direct product of its Sylow subgroups. The choice of N shows that G/N is an r-group for some prime r. But $|zN| \neq 1$ divides $q = p^i$. Consequently $r = p$ and G/N is a p-group. Therefore $G \in RF_{\{p\}}$.

   (b) Now $\tau(G) = \times_{p \in \pi} G_p$, where G_p is a p-group. Set $G_{p'} = \langle G_q : q \neq p \rangle$. If $z \in \tau(G) \backslash G_{p'}$, pick $N \triangleleft G$ maximal subject to $G_{p'} \leq N$ and $z \notin N$. As in the previous case G/N is a finite p-group. Also $G/\tau(G) \in RF_{\{p\}}$ by part (a). Hence $G/G_{p'} \in RF_{\{p\}}$. Clearly $\bigcap_p G_{p'} = \langle 1 \rangle$. Thus $G \cong G/\bigcap_p G_{p'} \in R(\bigcup_{p \in \pi} F_{\{p\}}) \subseteq RF_\pi$. $\square$

There are other residual properties known to hold in finitely generated nilpotent groups. For example there is the following result of Higman (1955). If G is such a group and if π is any infinite set of primes, then $\bigcap_{p \in \pi} G^p$ is finite; equivalently, if G is also torsion-free, then $\bigcap_{p \in \pi} G^p = \langle 1 \rangle$. For generalizations of this (and a proof) see Wehrfritz (1972). It is easy to see that this result of Higman's does not extend to polycyclic groups (just try the infinite dihedral group). The same applies to Gruenberg's Theorem, as the following theorem shows. (It also shows the necessity of the finite pieces in 2.11.)

2.17   Theorem (Seksenbaev 1965) Let π be an infinite set of primes. If G is a polycyclic-by-finite group with $G \in \bigcap_{p \in \pi} R\mathbf{F}_{\{p\}}$, then G is torsion-free and nilpotent.

To prove 2.17 we need the following special case of something known as Learner's Lemma. We will apply it with $\mathbf{X} = \mathbf{F}_{\{p\}}$.

2.18 If A is a maximal abelian normal subgroup of a group G and if $G \in R\mathbf{X}$, then $G/A \in RQ\mathbf{X}$.

Proof Set $B = \bigcap(AN : N \triangleleft G$ with $G/N \in \mathbf{X})$. Now AN/N is abelian, so $(AN)' \leq N$ and $B' \leq \bigcap_{G/N \in \mathbf{X}} N = \langle 1 \rangle$, since G is residually $\mathbf{X}$. Thus B is an abelian normal subgroup of G containing A. Consequently $B = A$. Clearly $G/AN \in Q\mathbf{X}$. Therefore $G/A \in RQ\mathbf{X}$.

Proof of 2.17 If $x \in H \in R\mathbf{F}_{\{p\}}$ with $|x|$ finite, then there is a normal subgroup N_i of H with H/N_i a p-group and $x^i \notin N_i$, this for each i with $1 \leq i < |x|$. Set $N = \bigcap_i N_i$. Then H/N is a p-group and $\langle x \rangle \cap N = \langle 1 \rangle$. Thus $\langle x \rangle$ is a p-group. This shows that any element of finite order in a residually (finite p-group) is a p-element. Therefore in 2.17 the group G is torsion-free.

Let A be a maximal abelian normal subgroup of G (A exists by 2.5 for example). By 2.18 we have $G/A \in \bigcap_{p \in \pi} R\mathbf{F}_{\{p\}}$. By induction on the Hirsch number G/A is nilpotent. Pick $B \leq A$ with B normal in G, G/B nilpotent and B of least Hirsch number ($=$ rank here) amongst subgroups of G with these properties. Clearly we may assume that $B \neq \langle 1 \rangle$.

Let $p \in \pi$. There exists a normal subgroup N of G of finite index with G/N a finite p-group and B not contained in N, that is, with $B > B \cap N$. Now G/N is nilpotent, say of class c, so $[B, {}_cG] \leq B \cap N$. Thus $[B, G](B \cap N) < B$. Also $B/(B \cap N)$ is a p-group, so the cyclic group of order p is an image of $B/[B, G]$. This is for all p in the infinite set π. Hence $B/[B, G]$ is infinite and $h(B) > h([B, G])$. Also G/B is nilpotent, so $G/[B, G]$ is nilpotent. This contradicts the choice of B, so $B = \langle 1 \rangle$ and G is nilpotent. $\square$

H is a *subnormal* subgroup of the group G (we write $H \triangleleft \triangleleft G$) if there exists finitely many subgroups H_i of G with

$$H = H_0 \triangleleft H_1 \triangleleft H_2 \triangleleft \cdots \triangleleft H_r = G.$$

For example, cf. 2.14(a), every subgroup of a nilpotent group G is subnormal in G, while not every subgroup of the finite metabelian group Sym(3) is subnormal.

2.19 Theorem (Kegel 1966) Let H be a subgroup of the polycyclic-by-finite group G. Then H is subnormal in G if and only if HN/N is subnormal in G/N for every normal subgroup N of finite index in G. The latter condition is equivalent to $HN \triangleleft \triangleleft G$ for all such N.

So again the finite images of G determine the property for the group itself. Notice that in any group G, if

$$N \leq H = H_0 \lhd H_1 \lhd H_2 \lhd \cdots \lhd H_r = G$$

with N normal in G, then

$$H/N = H_0/N \lhd H_1/N \lhd H_2/N \lhd \cdots \lhd H_r/N = G/N$$

and conversely, so the final claim is immediate. If G is polycyclic-by-finite and $H \leq G$, then $H = \bigcap_N HN$ by 2.9, see proof of 2.10, where N ranges over the normal subgroups of G of finite index. Thus it is immediate that H is normal in G if and only if all the HN are normal in G. A problem in 2.19 is that the subnormal chains for the HN in G may have unbounded length as N varies. If

$$G = \langle a, x, y : a^x = a^{-1}, a^y = a^2, xy = yx, x^2 = 1 \rangle,$$

then G is finitely generated and soluble (even metabelian). If $H = \langle a, x \rangle \leq G$, then H is not subnormal in G (since $[y^{-n}, a, {}_{n-1}x] \notin H$ for every positive integer n) but every HN for N a normal subgroup of G of finite index is normal in G ($a \in H$ and $A/\langle a \rangle$ for $A = \langle a^G \rangle$ is a Prüfer 2-group, so $A \leq HN \lhd G$). For further details see Lennox (1976).

Exercise Prove that this G is isomorphic to the rational matrix group

$$\left\langle \begin{pmatrix} 1 & 0 \\ 1 & 1 \end{pmatrix}, \begin{pmatrix} -1 & 0 \\ 0 & 1 \end{pmatrix}, \begin{pmatrix} 2 & 0 \\ 0 & 1 \end{pmatrix} \right\rangle.$$

Proof Clearly we may assume that G is infinite, so G has a non-trivial free-abelian normal subgroup A. By induction on the Hirsch number we may also suppose that AH/A is subnormal in G/A; consequently AH is now subnormal in G. It suffices to prove that H is subnormal in AH. Now A is normal in G, so $A \cap H$ is normal in H. Also A is abelian, so $A \cap H$ is normal in A. Therefore $A \cap H$ is normal in AH. Let $x \mapsto x^*$ denote the natural projection of AH onto $AH/(A \cap H)$. Then $(AH)^* = A^*]H^*$, the split extension of A^* by H^*.

Also by induction $A^n H$ is subnormal in G for $n = 1, 2, \ldots$, so $(A^n)^* H^*$ is subnormal in $A^* H^*$, say in $k(n)$ steps. Then $[A^* H^*, {}_{k(n)}(A^n)^* H^*] \leq (A^n)^* H^*$, after $k(n)$ applications of 1.4. Hence

$$[A^*, {}_{k(n)} H^*] \leq (A^n)^* H^* \cap A^* = (A^n)^* (H^* \cap A^*) = (A^n)^*.$$

Set $t = |\tau(A^*)|$ and $r = h(A^*)$. If p is a prime not dividing t, then $|A^*/(A^p)^*| = p^r$ and so $[A^*, {}_r H^*] \leq \bigcap_p (A^p)^*$, where p runs over all primes p not dividing t. This intersection is $\tau(A^*)$. Hence

$$[A^*, {}_{\max\{r,k(t)\}} H^*] \leq \tau(A^*) \cap (A^t)^* = \langle 1 \rangle.$$

Set $m = \max\{r, k(t)\}$. Then

$$H^* \lhd H^*[A^*, {}_{m-1}H^*] \lhd H^*[A^*, {}_{m-2}H^*] \lhd \cdots \lhd H^*A^*.$$

Therefore, since $A \cap H \leq H$, we have that H is subnormal in AH, which is subnormal in G. The proof is complete. $\square$

In fact 2.19 extends 2.8. To see this one needs the following.

2.20   If G is a polycyclic-by-finite group the following are equivalent.

(a) G is nilpotent.
(b) Every subgroup of G is subnormal in G.
(c) G has normalizer condition.

Proof Since $G \in$ **Max** it is easy to see that (b) and (c) are equivalent. Also (a) implies (c) by 2.14(a). Suppose (c). If G^* is any finite image of G, then G^* is a direct product of its p-subgroups and hence is nilpotent by 2.14(b). By 2.8 the group G is nilpotent; that is, (a) holds. $\square$

We summarize some further examples of the influence of the finite images of a polycyclic-by-finite group on the group itself.

2.21   Theorem (Lennox and Wilson 1977) Let H and K be subgroups of the polycyclic-by-finite group G. The following are equivalent.

(a) $HK = KH$.
(b) $H\phi K\phi = K\phi H\phi$ for all homomorphisms ϕ of G to finite groups.
(c) $HKN = KHN$ for all normal subgroups N of G of finite index.

Clearly (a) implies (b) and it is very easy to see that (b) and (c) are equivalent. Thus (b) implies (a) is the meat of the theorem. See Lennox and Wilson (1977, 1979) for proofs.

2.22   Theorem (Remeslennikov 1969; Formanek 1970, 1976) Let G be a polycyclic-by-finite group. If x and y are elements of G, then x and y are conjugate in G if and only if $x\phi$ and $y\phi$ are conjugate in $G\phi$ for every homomorphism ϕ of G to a finite group.

Apart from the original references, see Segal (1983), p. 59 for a proof. A group in which the conjugacy of elements is determined by the finite images of the group, as in 2.22, is said to be *conjugacy separable*. Behind 2.22 lies some substantial number theory, so its proof is not elementary. For the nilpotent case see 5.14 below and for the more recent result that the free product of two polycyclic-by-finite groups amalgamating a cyclic subgroup is conjugacy separable, see Ribes et al. (1998).

The reader may be beginning to wonder whether a polycyclic group is actually determined by its finite images. This is not the case. There exist non-isomorphic polycyclic groups (even nilpotent ones) with the same set of finite images. For example Baumslag (1974) shows that

$$\langle a, b | a^{25} = 1, a^b = a^6 \rangle \quad \text{and} \quad \langle a, b | a^{25} = 1, a^b = a^{11} \rangle$$

are not isomorphic but have the same set of finite images (these groups are in $\mathbf{C}_{25}\mathbf{C}_\infty$ and $\mathbf{C}_\infty(\mathbf{C}_{25}\mathbf{C}_5)$). However we do have the following very deep theorem of Grunewald et al. (1980). A full account of its proof is given in Segal (1983).

2.23   Theorem (Grunewald et al. 1980) The collection of all polycyclic-by-finite groups with a given set of finite images is the union of a finite set of isomorphism classes.

In other words, you cannot get an infinite set of pairwise non-isomorphic polycyclic-by-finite groups all having the same set of finite images.

The Profinite Topology

This gives another way of thinking about some of the theorems above. Let G be any group. Topologize G by taking the set of normal subgroups of G of finite index to be a basis of open neighborhoods of the identity. Thus a subset X of G is open by definition if X is a union of cosets of various normal subgroups of G of finite index. Check that this does impose a topology on G. (Main step: if M and N are normal subgroups of G of finite index and if $x, y \in G$, either $xM \cap yN = \emptyset$ or there exists z in $xM \cap yN$; in the latter case $xM \cap yN = zM \cap zN = z(M \cap N)$ and $M \cap N$ is a normal subgroup of G of finite index). A subset Y of G is closed if and only if $Y = \bigcap_N YN$, where N runs over the normal subgroups of G of finite index. In general set $\bigcap_N YN = Y^\wedge$. Clearly $Y^\wedge \supseteq Y$. Now if $G \backslash Y$ is open and $g \in G \backslash Y$, then $gN \subseteq G \backslash Y$ for some $N \triangleleft G$ of finite index and so $g \notin Y^\wedge$. Hence if Y is closed then $Y^\wedge = Y$. If $Y^\wedge = Y$ and $g \in G \backslash Y$, then $g \notin YN$ for some N. Thus $gN \subseteq G \backslash Y$. Consequently $G \backslash Y$ is open and Y is closed.

This topology is called the profinite topology of G. It makes G into a topological group; that is, $x \mapsto x^{-1}$ of G to G and $(x, y) \mapsto xy$ of $G \times G$ (with product topology) to G are both continuous. If H is an open subgroup of G, then $(G : H)$ is finite by definition of open set. If H is a subgroup of G of finite index, then $H_G = \bigcap_{g \in G} Hg$ is normal of finite index in G. Thus H_G is open and H as a union of cosets of H_G is also open. The profinite topology is the weakest topology making G into a topological group with every subgroup of finite index open.

2.24   Let H be a subgroup of the polycyclic-by-finite group G.

(a) H is closed in G in the profinite topology of G.
(b) If $K \leq H$ has finite index in H, then there exists $L \leq G$ of finite index in G such that $K = H \cap L$.
(c) The profinite topology of G induces on H the profinite topology of H.

Proof (a) This is just 2.9 rephrased.

(b) From (a) we have $K = \bigcap_F F$, where F ranges over all the subgroups of G of finite index containing K. Since $(H : K)$ is finite there exists finitely many of these F, say $F_1, F_2, \ldots, F_r$, with $K = H \cap F_1 \cap F_2 \cap \cdots \cap F_r$. Set $L = F_1 \cap F_2 \cap \cdots \cap F_r$.

(c) If N is normal in G of finite index, then $H \cap N$ is normal in H of finite index and hence $H \cap N$ is open in the profinite topology of H. Suppose now that N is a normal subgroup of H of finite index. By (b) there exists $L \leq G$ of finite index with $N = H \cap L$. Thus N is open in H in the topology induced on H by the profinite topology of G. The claim follows. $\qquad\square$

Let G be a polycyclic-by-finite group. Then 2.22 is equivalent to each conjugacy class of elements of G being closed in the profinite topology. In Lennox and Wilson (1979) is proved that if H and K are subgroups of G then the product set HK is closed in the profinite topology of G. Then 2.21 can easily be derived from this result.

Exercise If G is any group, prove that its profinite topology is Hausdorff if and only if G is residually finite.

Exercise If G and H are any groups taken with their profinite topologies, prove that any (group) homomorphism of G into H is continuous.

We complete this chapter by just mentioning a few further results about polycyclic groups that can be proved using the machinery we have available so far. Wehrfritz (1994) shows that the polycyclic group

$$\langle a, b, x \mid ab = ba, ax = xa^3b^4, bx = xa^2b^3 \rangle$$

contains no subgroup H of finite index with H/H' torsion-free (compare 2.7(d)). In the same paper it is shown that a finitely generated nilpotent-by-finite group has a nilpotent subgroup H of finite index such that H and H/H' are torsion-free.

Suppose ϕ is an endomorphism of the polycyclic-by-finite group G. Under any one of the following conditions ϕ is an automorphism of G.

(a) G is polycyclic and ϕ restricts to an automorphism of the centre $\zeta_1\eta_1(G)$ of the Fitting subgroup of G.
(b) ϕ is monic and restricts to an automorphism of $\zeta_1\eta_1(G)$.
(c) ϕ restricts to an automorphism of $\tau(G).\zeta_1\eta_1(G)$.
(d) $\tau(G)$ is soluble and ϕ restricts to an automorphism of $\zeta_1\eta_1(G)$.
(e) Farkas (1982) ϕ restricts to an automorphism of D, where $D/\tau(G) = \zeta_1\eta_1(G/\tau(G))$.

This subgroup D of the polycyclic-by-finite group G is called the Zalesskii subgroup of G. Clearly (a) here is a special case of (d). The above can be found in

Wehrfritz (2009a), a paper that extends ideas from Farkas (1982) and Wehrfritz (1983).

Nikolov and Segal (2007) prove the following. Let G be a polycyclic-by-finite group and M a normal subgroup of G. Then M is a direct factor of G if and only if MG^n/G^n is a direct factor of G/G^n for every positive integer n. They deduce that if G is isomorphic to the direct product of M and G/M, then M is a direct factor of G.

Exercise Let G be a polycyclic-by-finite group. Prove that G is nilpotent if and only if for each prime p there is a normal subgroup N_p of G with G/N_p a finite p-group such that $\bigcap_p N_p = \langle 1 \rangle$. (Hint: Slightly modify 2.18 and the proof of 2.17 for one direction and use Higman (1955) for the other.)

Chapter 3
Some Ring Theory

Noetherian Rings

All our rings will have an identity and all our modules will be unital. Our modules
will sometimes be right, sometimes be left and sometimes have actions on both sides
(e.g. bimodules). The following is an analogue of 2.3.

3.1 Let R be a ring and M a R-module (left or right). The following are equivalent.

(a) Every submodule of M is finitely generated.
(b) Every ascending chain $M_1 \le M_2 \le \cdots \le M_i \le \cdots$ of submodules of M contains only finitely many distinct members.
(c) Every non-empty set S of submodules of M has a maximal member.

Proof (a) implies (b). $N = \bigcup_i M_i$, where $i = 1, 2, \ldots$, is a submodule of M and
hence is finitely generated. Thus $N = M_n$ for some n and (b) follows.
  (b) implies (c). If S has no maximal member we can pick $M_1, M_2, \ldots$ in S,
(using the axiom of choice) with $M_1 < M_2 < \cdots < M_i < \cdots$. This contradicts (b).
  (c) implies (a). Let S denote the set of finitely generated submodules of the submodule N of M. By hypothesis S has a maximal member X. Then X is finitely
generated. If $x \in N$, then $\langle X, x \rangle \in S$, so $x \in X$ by the maximality of X and $N = X$,
which is finitely generated. $\qquad\square$

An R-module satisfying the conditions of 3.1 is said to be *Noetherian*. The following is more or less immediate from 3.1(a)

3.2 Let N be a submodule of the R-module M. Then M is Noetherian if and only
if N and M/N are Noetherian.
  The ring R is left (resp. right) Noetherian if R is Noetherian as left (resp. right)
R-module. R is a *Noetherian* ring if R is both left and right Noetherian. We frequently just state and prove results for right modules, the left version being left to
the reader.

B.A.F. Wehrfritz, *Group and Ring Theoretic Properties of Polycyclic Groups*, 29
Algebra and Applications 10,
DOI 10.1007/978-1-84882-941-1_3, © Springer-Verlag London Limited 2009

3.3 Let R be a right Noetherian ring and M a finitely generated right R-module. Then M is Noetherian.

Proof Suppose $M = x_1 R + x_2 R + \cdots + x_n R$. By induction on n we may assume that $M/x_1 R$ is Noetherian. The map $r \mapsto x_1 r$ is a homomorphism of R_R onto $x_1 R$. By 3.2 the latter is Noetherian; consequently so is M by 3.2 again.   □

Exercise If R is a commutative ring and M is a Noetherian R-module, prove that $R/\operatorname{Ann}_R M$ is a Noetherian ring. Here $\operatorname{Ann}_R M = \{r \in R : Mr = 0\}$.

Group Rings

Let J be a ring and G a group with $J \cap G = \{1\}$. (This is really just for notational convenience; if $J \cap G \neq \{1\}$ replace G by an isomorphic copy of itself so that we do get this equality.)

Let JG denote the free left J-module with basis G. That is, each element of JG has a unique representation in the form $\sum_{g \in G} \alpha_g g$, where the α_g lie in J and almost all are zero. Addition is componentwise; viz.

$$\sum \alpha_g g + \sum \beta_g g = \sum (\alpha_g + \beta_g) g.$$

Since $J \cap G = \{1\}$, both $J = J. 1$ and $G = 1. G$ are subsets of JG.

Define a multiplication on JG by

$$\left(\sum \alpha_g g \right) \left(\sum \beta_h h \right) = \sum_{g,h \in G} (\alpha_g \beta_h)(gh).$$

This is well defined since almost all the $\alpha_g \beta_h$ are zero. One easily checks that JG becomes a ring with J a subring of JG and G a subgroup of the group of units of JG. For example,

$$\left(\left(\sum \alpha_g g \right) \left(\sum \beta_g g \right) \right) \left(\sum \gamma_g g \right) = \sum_{g,h,k} (\alpha_g \beta_h) \gamma_k (gh) k$$

$$= \left(\sum \alpha_g g \right) \left(\left(\sum \beta_g g \right) \left(\sum \gamma_g g \right) \right).$$

Note that $\alpha g = g \alpha$ for all α in J and g in G. Then $\mathbf{Z} G$ is called the group ring of G (recall $\mathbf{Z}$ denotes the integers), while JG is called the group ring of G over J.

If R is any ring and if $\phi : G \to R$ is a homomorphism of G into the group of units of R, then $\phi^\sim : \mathbf{Z} G \to R$ given by $\left(\sum \alpha_g g \right) \phi^\sim = \sum \alpha_g (g \phi)$ is a ring homomorphism. In particular if ϕ is a group homomorphism of G to say H, then $\phi^\sim$ is a homomorphism of the group ring $\mathbf{Z} G$ to the group ring $\mathbf{Z} H$. In general ϕ defines a ring homomorphism of JG to JH.

The map $\varepsilon : JG \to J$ given by $\sum \alpha_g g \mapsto \sum \alpha_g$ is a ring homomorphism of JG onto J called the *augmentation map*. Its kernel $\ker \varepsilon$ is called the *augmentation ideal* of JG. It depends on $G \subseteq JG$ rather that just JG as a ring. We write $\mathbf{g}$ for the augmentation ideal of JG, the J usually being understood. Thus $\mathbf{h}$ denotes the augmentation ideal of JH, $\mathbf{g}_1$ the augmentation ideal of G_1 etc. One has to be slightly careful with this notation. An ideal of a ring R we normally denote by a bold letter, say $\mathbf{k}$. Only if a group ring JK has already been introduced in an argument would $\mathbf{k}$ denote its augmentation ideal without further comment.

3.4   Let J be a ring, G a group and JG the corresponding group ring.

(a) $\mathbf{g} = \bigoplus_{g \in G \backslash \langle 1 \rangle} (g - 1)J = \bigoplus_{g \in G \backslash \langle 1 \rangle} J(g - 1)$ and $JG = J \oplus \mathbf{g}$.
(b) If $G = \langle X \rangle$, then $\mathbf{g} = \sum_{x \in X} (x - 1)JG = \sum_{x \in X} JG(x - 1)$.

Proof (a) $JG = \bigoplus_{g \in G} Jg$. Thus $\mathbf{g} \supseteq \sum_{g \in G \backslash \langle 1 \rangle} J(g - 1) = \bigoplus_{g \in G \backslash \langle 1 \rangle} J(g - 1)$. If $x = \sum \alpha_g g \in \mathbf{g}$, then $\sum \alpha_g = 0$, so $x = \sum \alpha_g g - \sum \alpha_g = \sum \alpha_g (g - 1)$. Thus the above containment is an equality. The right J-module version follows immediately. It also follows that $JG = J + \mathbf{g}$ and since ε is monic on J we obtain $JG = J \oplus \mathbf{g}$.

(b) If $x, y \in G$, then $xy - 1 = (x - 1)y + (y - 1)$ and $x^{-1} - 1 = -(x - 1)x^{-1}$. It follows easily that $\mathbf{g} \leq \sum_{x \in X} (x - 1)JG \leq \mathbf{g}$. The second equality follows similarly.   □

Modules over Groups

Let G be a group and M an additive abelian group. Suppose we have a map of $M \times G$ into M, denoted by $(m, g) \mapsto mg$, such that

$$(m_1 + m_2)g = m_1 g + m_2 g, (mg_1)g_2 = m(g_1 g_2) \quad \text{and} \quad m1 = m$$

for all m, m_1, m_2 in M and g, g_1, g_2 in G. We say that this information constitutes a right G-module M. The first identity here says that $g \in G$ defines an endomorphism ϕ_g of the abelian group M. The other two imply that $(mg)g^{-1} = m(gg^{-1}) = m$, so ϕ_g is actually an automorphism of M. The second identity then says that $\phi : g \mapsto \phi_g$ is a (group) homomorphism of G into the automorphism group $\text{Aut}_{\mathbf{Z}} M$ of M. As we have seen above ϕ extends to a ring homomorphism $\phi^{\sim}$ of $\mathbf{Z}G$ into $\text{End}_{\mathbf{Z}} M$. Define an action of $\mathbf{Z}G$ on M by $mx = m(x\phi^{\sim})$ for $m \in M$ and $x \in \mathbf{Z}G$. Then M becomes a right $\mathbf{Z}G$-module. To summarize, if M is a right G-module, then M becomes a right $\mathbf{Z}G$-module by setting $m(\sum \alpha_g g) = \sum \alpha_g mg$. Conversely if M is a right $\mathbf{Z}G$-module, then M becomes a right G-module simply by restricting the acting set from $\mathbf{Z}G$ to G. There is a corresponding notion of left G-module that relates to left $\mathbf{Z}G$-modules in a similar way.

It is often convenient to write G-modules multiplicatively. Then M is a multiplicative G-module if M is an abelian group written multiplicatively and carrying a map $(m, g) \mapsto m^g$ of $M \times G$ to M such that $(m_1 m_2)^g = m_1^g m_2^g$, $m^{g_1 g_2} = (m^{g_1})^{g_2}$ and $m^1 = m$, where m, m_1, m_2, g, g_1 and g_2 are as above.

Exercise Let A be an abelian normal subgroup of the group E and set $G = E/A$. Check that A becomes a right E-module with action given by $a^e = e^{-1}ae$ and becomes a right G-module with action given by $a^{eA} = e^{-1}ae$. Show that the submodules of A are exactly the normal subgroups of E contained in A.

More generally the following is easily seen to be true. Let $K' \leq H \leq K \leq L$ be subgroups of a group L. Suppose $S \leq N_L(H) \cap N_L(K)$ and suppose T is a normal subgroup of S in $C_S(K/H) = \{s \in S : [K, s] \subseteq H\}$. Set $M = K/H$ and $G = S/T$. Then M becomes a well-defined G-module with action given by $(kH)^{(sT)} = k^s H$. If $K = A$, $H = \langle 1 \rangle$ and $S = E$ we obtain the first case in the exercise. If also $T = A$ we obtain the second. We can also take $T = C_E(A)$ and obtain that M is an $E/C_E(A)$-module and a $G/C_G(A)$-module in a natural way.

We wish to study a finitely generated, abelian-by-polycyclic-by-finite group E.

Exercise $A(PF) = (AP)F$, so this is not ambiguous.

Choose A abelian and normal in E with $G = E/A$ polycyclic-by-finite. Our strategy will be as follows. Given G we have information about the ring $\mathbf{Z}G$. This gives information about $\mathbf{Z}G$-modules and hence about the $\mathbf{Z}G$-module A. Since we now 'know' about A and G, we know something about E. First we must study the group ring $\mathbf{Z}G$ of a polycyclic-by-finite group G.

Noetherian Group Rings

3.5 Let J be a ring and G a group. If JG is right Noetherian, then J is right Noetherian and $G \in \mathbf{Max}$.

Proof Since $J \cong JG/\mathbf{g}$, so J is right Noetherian. Let H be a subgroup of G. We claim that

3.5.1 $(1 + \mathbf{h}JG) \cap G = H$.

If 3.5.1 is true and if $H_1 < H_2 < \cdots$ is a chain of distinct subgroups of G then $\mathbf{h}_1 JG \leq \mathbf{h}_2 JG \leq \cdots$ is a chain of distinct right submodules of JG. Thus if JG is right Noetherian, then G has the maximal condition on subgroups; that is, $G \in \mathbf{Max}$.

It remains to prove 3.5.1. If $h \in H$, then $h = 1 + (h - 1) \in (1 + \mathbf{h}JG) \cap G$. Let T denote a right transversal of H to G, so G is the disjoint union of the cosets Ht for $t \in T$. Then

$$JG = \bigoplus Jg = \bigoplus_{t \in T} \left(\bigoplus_{h \in H} Jh \right) t = \oplus_{t \in T} JHt.$$

Hence $\mathbf{h}JG = \bigoplus_{t \in T} \mathbf{h}t$. If $g \in (1 + \mathbf{h}JG) \cap G$, then $g = ht$ for some unique h in H and t in T. Thus $ht - 1 \in \bigoplus_{u \in T} \mathbf{h}u$. Also $h \notin \mathbf{h}$ (since the augmentation map maps h to 1 and $\mathbf{h}$ to $\{0\}$). Therefore $t \in H$ and $g = ht \in H$. Result 3.5.1 follows. $\square$

3.6 Let J be a commutative ring and G any group. If JG is right (resp. left) Noetherian, then JG is left (resp. right) Noetherian and hence is Noetherian.

We are especially interested in the case where $J = \mathbf{Z}$.

Proof The map $\iota : JG \to JG$ given by $\sum \alpha_g g \mapsto \sum \alpha_g g^{-1}$ satisfies $\iota^2 = 1$ and is an anti-automorphism; that is, it is an additive automorphism satisfying $(xy)\iota = (y\iota)(x\iota)$ for all x and y in JG. Then M is a right JG-submodule of JG if and only if $M\iota$ is a left JG-submodule of JG. Also $N \leq M$ if and only if $N\iota \leq M\iota$. The result follows. $\qquad\square$

3.7 Theorem (Hall 1954) If J is a right Noetherian ring and if G is a polycyclic-by-finite group, then the group ring JG is right Noetherian.

Thus if G is a polycyclic-by-finite group, then $\mathbf{Z}G$ is a Noetherian ring. Hence by 3.5 and 3.7 we have $\mathbf{PF} \subseteq \{\text{groups } G : \mathbf{Z}G \text{ is Noetherian}\} \subseteq \mathbf{Max}$. Also, as mentioned in Chap. 2, it is known that $\mathbf{PF} \neq \mathbf{Max}$. As far as I know it is still not known whether $\mathbf{Max} = \{G : \mathbf{Z}G \text{ is Noetherian}\}$ or whether $\mathbf{PF} = \{G : \mathbf{Z}G \text{ is Noetherian}\}$. At least one is false and quite likely both are. We need a couple of lemmas to prove 3.7.

3.7.1 Lemma (McConnell 1968) Let R be a ring, S a right Noetherian subring of R and x an element of R such that $S + xS = S + Sx$ and $R = \langle S, x \rangle$, the subring of R generated by S and x. Then R is right Noetherian.

Proof Since $xS \subseteq S + Sx$, an easy induction yields that

$$x^n S \subseteq S + Sx + Sx^2 + \cdots + Sx^n \quad \text{for all } n \geq 1.$$

Thus using symmetry we obtain that

$$S + xS + x^2 S + \cdots + x^n S = S + Sx + Sx^2 + \cdots + Sx^n \quad \text{for all } n \geq 1$$

and that $R = \sum_{n \geq 0} Sx^n$. Let $s' \in S$ and $n \geq 0$. Then there exists $s'' \in S$ with

$$x^n s'' = (\text{sum of terms in } Sx^i \text{ for } i < n) + s'x^n. \qquad (*)$$

Let I be any right ideal ($=$ right R-submodule) of R and let I^* denote the set of all elements s of S for which there exists $n \geq 0$ and $r \in I$ with

$$r = (\text{sum of terms in } Sx^i \text{ for } i < n) + sx^n.$$

We claim that I^* is a right ideal of S. Certainly it is not empty. If also we have

$$r_1 = (\text{sum of terms in } Sx^i \text{ for } i < m) + s_1 x^m \in I,$$

then

$$rx^m + r_1 x^n = \text{(terms of lower degree)} + (s + s_1)x^{m+n} \in I.$$

Thus $I^* + I^* \subseteq I^*$. If also $s' \in S$, let s'' be as in $(*)$. Then

$$rs'' = \text{(terms of lower degree)} + sx^n s'' = \text{(terms of lower degree)} + ss'x^n \in I,$$

so $ss' \in I^*$ and hence $I^*s' \subseteq I^*$. Therefore I^* is a right ideal of S as claimed.

Since S is right Noetherian, $I^* = x_1^* S + x_2^* S + \cdots + x_m^* S$ for some elements $x_1^*, \ldots, x_m^*$ of I^*. By definition there exists for each i an element x_i of I and integer $n(i)$ with

$$x_i = \text{(terms of lower degree)} + x_i^* x^{n(i)}.$$

Set $n^* = \max\{n(1), n(2), \ldots, n(m)\}$. Then $N_0 = \sum_{0 \le i \le n^*} Sx^i = \sum_{0 \le i \le n^*} x^i S$ is finitely generated as right S-module and consequently, by the Noetherian condition on S, so is $N = N_0 \cap I$. Set $J = \sum_{1 \le i \le m} x_i R + NR$ (actually $J = NR$, but this is only relevant if one wishes to choose a 'small' generating set). Then J is a finitely generated right ideal of R and by construction $J \subseteq I$. If we can show that $I = J$, then I is finitely generated and R is right Noetherian as required.

Assuming $I \ne J$, pick $r = \text{(terms of lower degree)} + sx^n \in I \setminus J$ with n minimal. Then $s \in I^*$, so $s = x_1^* s_1' + \cdots + x_m^* s_m'$ for some s_i' in S. If $n \le n^*$, then $r \in N \subseteq J$, so $n > n^*$. Let

$$f = x_1 x^{n-n(1)} s_1'' + \cdots + x_m x^{n-n(m)} s_m''$$

where s_i'' are chosen by $(*)$ so that

$$x^n s_i'' = \text{(terms of lower degree)} + s_i' x^n.$$

Then $f = \text{(terms of lower degree)} + (x_1^* s_1' + \cdots + x_m^* s_m')x^n = \text{(terms of lower degree)} + sx^n$. Consequently $r - f \in \sum_{0 \le i < n} Sx^i$. Now each $x_i \in J$, so $f \in J \subseteq I$ and therefore $r - f \in I$. By the choice of r we have $r - f \in J$ and so $r \in f + J \subseteq J$. This contradiction completes the proof that $I = J$. □

3.7.2 Lemma (Hall 1954) Let R be a ring, S a right Noetherian subring of R and x a unit of R such that $xS = Sx$ (equivalently $x^{-1}Sx = S$; that is, x 'normalizes' S) and $R = \langle S, x, x^{-1} \rangle$. Then R is right Noetherian.

Proof By 3.7.1 the subring $T = \langle S, x \rangle$ of R is right Noetherian. Clearly $x^{-1}S = Sx^{-1}$, so $x^{-1}T = Tx^{-1}$ and $R = \langle T, x^{-1} \rangle$. Thus 3.7.2 follows from a second application of 3.7.1. □

Proof of 3.7 The group G has a series of finite length whose factors are cyclic or finite. We induct on the length of such a series. Thus we may assume there is an H normal in G with G/H cyclic or finite and JH right Noetherian.

Suppose G/H is infinite cyclic. Then $G = H\langle x \rangle$ for some x in G. Clearly $JG = \bigoplus_{i \in \mathbf{Z}} JHx^i$ and $x^{-1}(JH)x = JH$. Hence JG is right Noetherian by 3.7.2, where we take JG for R and JH for S. Now assume G/H is finite. If T is a transversal of H to G (meaning G is the disjoint union of the cosets $Ht = tH$ for $t \in T$), then $JG = \bigoplus_{t \in T} t JH$. Thus JG is finitely generated as right JH-module. By 3.3 the ring JG is Noetherian as right JH-module and hence also as right JG-module. The proof is complete. $\qquad\qquad\qquad\qquad\qquad\qquad\qquad\qquad\qquad\qquad\qquad\qquad\qquad$ $\square$

The proof above also proves the following generalization of both 3.7.2 and 3.7.

3.7.3 (Hall 1954) The ring R is generated as a ring by its right Noetherian subring J and the subgroup G of its group of units. Assume that G normalizes J and is polycyclic-by-finite. Then R is right Noetherian.

Applications

Let G be a polycyclic-by-finite group. Then $\mathbf{Z}G$ is Noetherian by 3.7 and so finitely generated G-modules are Noetherian. The last link in our planned chain is to use information on G-modules to study the class $\mathbf{G} \cap \mathbf{APF}$. How do we fill in the last link in this chain?

Wreath Products. Let G and H be any groups.
The Standard Wreath Product HwrG. Let $B = \times_{g \in G} H_g$ be the direct product of copies of H indexed by the elements of G. For each $g \in G$ let $h \mapsto h_g$ be a fixed isomorphism of H to H_g. Now G acts on B by permuting the copies of H via its action on the suffices, meaning that if $g, x \in G$ and $h \in H$, then $(h_g)^x = h_{gx}$. Set $HwrG = G[B$, the split extension of B by G. The subgroup B is called the base group of this wreath product.
The Complete Wreath product HWrG. Let $B^\wedge = \prod_{g \in G} H_g$, the cartesian product of the copies H_g of H. Again G acts on $B^\wedge$ by permuting the suffices and we set $HWrG = G[B^\wedge$, the split extension of $B^\wedge$ by G. Again $B^\wedge$ is called the base group of the wreath product. In many ways the above is the easiest way to picture a wreath product. It is frequently not the easiest way to compute with it.

Now $\mathrm{Map}(G, H)$ and $\prod_{g \in G} H_g$ are isomorphic for example via $\beta \mapsto (\beta(g)_g)$. Thus instead set $B^\wedge = \mathrm{Map}(G, H)$, with the maps written on the left. Define an action of G on $B^\wedge$ as follows. If $x, y \in G$ and $\beta \in B^\wedge$, set $\beta^y : x \mapsto \beta(yx)$. Now set $HWrG$ equal to the split extension of $B^\wedge$ by G. These two definitions of $HWrG$ yield isomorphic groups (see exercises below). Also

$$B \cong \{\beta \in B^\wedge : \operatorname{supp} \beta = \{x \in G : \beta(x) \neq 1\} \text{ is finite}\}.$$

3.8 Theorem (Kaluzhnin-Krasner Theorem) Let E be a group and H a normal subgroup of E and set $G = E/H$. Then there exists an embedding ϕ of E into $W^\wedge =$

$HWrG$ such that $H\phi \leq B^\wedge$ and $\pi = \phi\pi$, where $\pi : E \to G$ and $\pi^\wedge : W^\wedge \to G$ are the natural projections.

Thus we obtain the following commutative diagram.

$$
\begin{array}{ccccccccc}
1 & \longrightarrow & H & \longrightarrow & E & \longrightarrow & G & \longrightarrow & 1 \\
 & & \downarrow & & \downarrow{\scriptstyle\phi} & & \parallel{\scriptstyle=} & & \\
1 & \longrightarrow & B^\wedge & \longrightarrow & W^\wedge & \longrightarrow & G & \longrightarrow & 1
\end{array}
$$

Proof Let T be a transversal of H to E. Label the elements of T so that $T = \{t_x : x \in G\}$ where $t_x\pi = x$ for each x in G.

If $e \in E$, then $et_x = t_{x(e)}h_{e,x}$ for some unique $x(e)$ in G and $h_{e,x}$ in H. Apply π. We obtain $e\pi.x = x(e)$. Define $\beta_e \in B^\wedge$ by $\beta_e : x \mapsto h_{e,x}$. Then define $\phi : E \to W^\wedge = GB^\wedge$ by $\phi : e \mapsto (e\pi).\beta_e \in GB^\wedge = W^\wedge$.

ϕ is a homomorphism, for if $e, f \in E$, then $ef.t_x = et_{f\pi.x}h_{f,x} = t_{e\pi.f\pi.x}h_{e,f\pi.x}h_{f,x}$. Now $(ef)\phi = (ef)\pi.\beta_{ef} = e\pi.f\pi.\beta_{ef}$ and $(e\phi)(f\phi) = e\pi.\beta_e.f\pi.\beta_f = e\pi.f\pi.\beta_e^{f\pi}.\beta_f$. Also if $x \in G$, then $\beta_{ef} : x \mapsto h_{e,f\pi.x}.h_{f,x}$ and $\beta_e^{f\pi}.\beta_f : x \mapsto h_{e,f\pi.x}.h_{f,x}$. Therefore ϕ is a homomorphism.

ϕ is one-to-one, for if $e\phi = 1$, then $e\pi = 1$ and $\beta_e = 1$. The first equation yields that $e \in H$. Now H is normal in G, so $et_x = t_x(t_x^{-1}et_x) = t_xh_{e,x} = t_x(\beta_e(x))$. But $\beta_e = 1$. Therefore $e = 1$ and ϕ is one-to-one.

Finally $H\pi = \langle 1 \rangle$, so $H\phi \leq B^\wedge$. Also $e\phi\pi^\wedge = e\pi$ by definition, so $\phi\pi^\wedge = \pi$. The proof is complete.   □

Exercise There are two obvious right actions of G on $C = \prod_{g \in G} H_g$ and two on $D = \mathrm{Map}(G, H)$. They all lead to the same wreath product. The following exercises will lead you through this. The two actions on the cartesian product are given for $g, x \in G$ and $h(g) \in H$ by

$$
(h(g)_g)\rho_x = ((h(g))_g)^x = (h(g)_{gx}) = (h(gx^{-1})_g),
$$

the one used above, and $(h(g)_g)\lambda_x = (h(g)_{x^{-1}g}) = (h(xg)_g)$.

Prove that $\theta : (h(g)_g) \mapsto (h(g)_{g^{-1}})$ is a bijection of the cartesian product interchanging these two actions of G.

If we let G act on D via $\beta^x(g) = \beta(xg)$ prove that $\phi : \beta \mapsto (\beta(g)_{g^{-1}})$ is a bijection of D to C between this action of G on D and the ρ_x action of G on C. On the other hand $\psi : \beta \mapsto (\beta(g)_g)$ is a bijection of D to C between this action of G on D and the λ_x action on C. There is also the action $\beta\lambda_x(g) = \beta(gx^{-1})$ of G on D. Then ϕ sends this action to the λ_x action on C and ψ sends it to the ρ_x action. It is very easy to check all this. It follows that any one of these actions can be used to define the complete wreath product.

Suppose we have a finitely generated group E and an abelian normal subgroup A of E. Set $G = E/A$ and let $\pi : E \mapsto G$ be the natural projection. Let $\phi : E \to W^\wedge =$

$AWrG = G[B^\wedge$ be the embedding constructed in 3.8. Now $G = E\pi$ and $E\phi$ are both finitely generated, so $E_1 = \langle G, E\phi \rangle \leq W^\wedge$ is also finitely generated. Further $B^\wedge$ is abelian.

Set $A_1 = E_1 \cap B^\wedge$. Then A_1 is an abelian normal subgroup of E_1. Also $E_1 = E_1 \cap GB^\wedge = G(E_1 \cap B^\wedge)$. Thus E_1 is the split extension $G[A_1$ of A_1 by G. Further $A\phi \leq A_1$ and we have the following diagram.

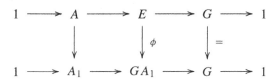

E_1 is finitely generated, say by $g_1a_1, g_2a_2, \ldots, g_na_n$, where the g_i lie in G and the a_i lie in A_1. Then if $w(x_1, x_2, \ldots, x_n)$ is any word in the n exhibited variables, then $w(g_1a_1, g_2a_2, \ldots, g_na_n) = w(g_1, g_2, \ldots, g_n)y$, where y is some word in the a_i and their conjugates under G. Hence $a_1, a_2, \ldots, a_n$ generate A_1 as a normal subgroup of E_1; that is, they generate A_1 as a G-module. Thus if we know about the group G and about finitely generated G-modules, then we certainly know about the split extensions GA_1, so we know something about the original group $E \cong E\phi \leq GA_1$. The above and 3.3 and 3.1 yield the following.

3.9   Corollary Let E be a finitely generated group and A an abelian normal subgroup of E, and set $G = E/A$. Then as G-module A is isomorphic to a submodule of a finitely generated G-module (namely the A_1 above.) If also $\mathbf{Z}G$ is Noetherian (e.g. if G is polycyclic-by-finite), then A is finitely G-generated (meaning A is finitely generated as a G-module).

3.10   Corollary (Hall 1954) Let G be a finitely generated, abelian-by-polycyclic-by-finite group. Then E satisfies max-n, the maximal condition on normal subgroups.

Thus, for example, finitely generated metabelian groups satisfy max-n.

Proof There exists A abelian and normal in E with $G = E/A$ polycyclic-by-finite. Then A is Noetherian as G-module by 3.9 and thus E has the maximal condition on those of its normal subgroups contained in A. Also E/A has max by 2.5.

Let $N_1 \leq N_2 \leq \cdots \leq N_i \leq \cdots$ be an ascending chain of normal subgroups of E. Then $\{A \cap N_i : i = 1, 2, \ldots\}$ is finite, so by deleting the first few terms we may assume that all the $A \cap N_i$ are equal. In the same way we may assume that all the N_iA/A are equal. Then by the following lemma $N_1 = N_i$ for all $i \geq 1$ and the claim will follow.   □

3.10.1   Let E be a group, H a normal subgroup of E and $X \leq Y$ subgroups of E with $XH = YH$ and $X \cap H = Y \cap H$. Then $X = Y$.

Proof For $Y = Y \cap YH = Y \cap XH = X(Y \cap H) = X(X \cap H) = X$.   □

Exercise Use 3.10.1 to give a second proof of $P\mathbf{Max} = \mathbf{Max}$, see 2.4. Also use 3.10.1 to prove that $P\mathbf{Min} = \mathbf{Min}$. (A group lies in $\mathbf{Min}$ if it satisfies the minimal condition on subgroups.)

We can also use 3.9 to give a third proof of $\mathbf{G} \cap \mathbf{N} \subseteq \mathbf{P}$, see 2.13. For let E be a finitely generated nilpotent group and set $A = \zeta_1(E)$. We induct on the class of E, so we may assume that E/A is polycyclic. Then A is finitely generated as E-module by 3.9. But A is central, so this says that A is a finitely generated abelian group. Therefore E is polycyclic.

3.11 (Hall 1954) A soluble-by-finite group E with max-n, the maximal condition on normal subgroups, is finitely generated. Thus

$$\mathbf{G} \cap \mathbf{APF} \subseteq \mathbf{SF} \cap \mathbf{Max\text{-}n} \subseteq \mathbf{G} \cap \mathbf{SF}.$$

Proof By induction on the derived length of a soluble normal subgroup of E of finite index (and the triviality $\mathbf{F} \subseteq \mathbf{G}$) we may assume that there is an abelian normal subgroup A of E with E/A finitely generated. Then there exists finitely many elements $x_1, x_2, \ldots, x_m$ of E such that $E = \langle x_1, x_2, \ldots, x_m \rangle A$. Since E satisfies the maximal condition on normal subgroups, A is finitely generated as a normal subgroup of E, say by $a_1, a_2, \ldots, a_n$. Thus $A = \langle a_1^E, \ldots, a_n^E \rangle$, where for any a in A

$$a^E = \{a^e : e \in E\} = \{a^x : x \in \langle x_1, x_2, \ldots, x_m \rangle\},$$

the latter since A is abelian. Therefore

$$A \leq \langle a_1, a_2, \ldots, a_n, x_1, x_2, \ldots, x_m \rangle \quad \text{and} \quad E = \langle a_1, a_2, \ldots, a_n, x_1, x_2, \ldots, x_m \rangle.$$

Thus E is finitely generated.

The above proves the second containment of the final claim; the first containment follows from 3.10.   □

Example The subgroup G of $GL(3, \mathbf{Q})$ generated by the rational three matrices

$$\begin{pmatrix} 1 & 0 & 0 \\ 1 & 1 & 0 \\ 0 & 0 & 1 \end{pmatrix}, \quad \begin{pmatrix} 1 & 0 & 0 \\ 0 & 1 & 0 \\ 0 & 1 & 1 \end{pmatrix}, \quad \begin{pmatrix} 1 & 0 & 0 \\ 0 & 2 & 0 \\ 0 & 0 & 1 \end{pmatrix}$$

is finitely generated and soluble. Its centre is isomorphic additively to $\mathbf{Z}[1/2]$, which is not $\mathbf{Z}$-Noetherian, so G does not satisfy the maximal condition on normal subgroups. Another such example is $W = Cwr(CwrC)$, where C is infinite cyclic. Clearly W is finitely generated and soluble. The second base group here is isomorphic to the group ring $R = \mathbf{Z}(CwrC)$. If W has max-n, then R is Noetherian and $CwrC$ is polycyclic (by 3.5). This is clearly false, so W does not satisfy max-n.

If instead we set $W = (C \, wr \, C) \, wr \, C$, C as above, then W is soluble with max-n but does not lie in $\mathbf{G} \cap \mathbf{APF}$. Clearly W is soluble. If the right hand C here is generated by x and the middle C is generated by y, then we have normal subgroups $A \leq B$ of W with $G = \langle x \rangle B = \langle x, y \rangle A$, $B/A \cong \mathbf{Z}[x, x^{-1}]$, where x operates by multiplication and $A \cong \mathbf{Z}[x, x^{-1}, y, y^{-1}]$, where x and y also operate by multiplication. Then A and B/A are Noetherian as W-modules and so W satisfies max-n. If W is in $\mathbf{G} \cap \mathbf{APF}$, one deduces that A is the Fitting subgroup of W and that $C \, wr \, C$ is polycyclic, which is clearly false.

Thus the containments in 3.11 are both strict, a fact also due to (Hall 1954).

Many results about finitely generated soluble-by-finite groups have been proved using the sort of techniques above. Some, especially those involving polycyclic groups, we shall see as we go along. The following we will not be covering, so we mention it now without proof.

3.12 Theorem (Lennox and Roseblade 1980) A soluble group G is polycyclic if $G = HK$, where H and K are both polycyclic subgroups of G.

Exercise (Gruenberg 1973) Let G be a group, H a normal subgroup of G and x an element of G that is central modulo H. Set $K = \langle x \rangle H$, $S = \mathbf{Z}H$ and $T = \mathbf{Z}K$. Prove that S satisfies the maximal condition on (2-sided) ideals normalized by G if and only if T does.

Exercise (Gruenberg 1973) If J is a ring and G a group such that JG satisfies the maximal condition on ideals, prove that G satisfies max-n, the maximal condition on normal subgroups.

Chapter 4
Soluble Linear Groups

Let F be a field and n a positive integer. $GL(n, F)$ denotes the multiplicative group of all n by n invertible matrices with entries in F. By definition, a *linear* group is a subgroup of $GL(n, F)$ for some n and F. Warning: a linear group is more that just a group; it is a group together with a particular embedding into some selected $GL(n, F)$. For example, working over the complex numbers the groups

$$X = \left\langle \begin{pmatrix} 1 & 0 \\ 1 & 1 \end{pmatrix} \right\rangle \quad \text{and} \quad Y = \left\langle \begin{pmatrix} 2 & 0 \\ 0 & 1 \end{pmatrix} \right\rangle$$

are isomorphic as groups, both being infinite cyclic, but have very different properties as linear groups. If R is a ring (with an identity as always), then $GL(n, R)$ denotes the obvious thing, namely the group of n by n invertible matrices over the ring R, but its subgroups will not be called linear groups unless, of course, R is a (commutative) integral domain.

There are other canonical linear groups that we need to refer to. $D(n, F) = \{(a_{ij}) \in GL(n, F) : a_{ij} = 0 \text{ whenever } i \neq j\}$ is the diagonal group. $Tr(n, F) = \{(a_{ij}) \in GL(n, F) : a_{ij} = 0 \text{ whenever } i < j\}$ is the (lower) triangular group. $Tr_1(n, F) = \{(a_{ij}) \in Tr(n, F) : a_{ii} = 1 \text{ for all } i\}$ is the lower unitriangular group. $Tr_1(n, F) = \{(a_{ij}) \in GL(n, F) : a_{ij} = 0 \text{ whenever } i > j \text{ and } a_{ii} = 1 \text{ for all } i\}$ is the upper unitriangular group. For example, $Tr(3, F)$ consists of all invertible matrices of the type A below while $Tr_1(3, F)$ consists of all matrices of type B:

$$A = \begin{pmatrix} * & 0 & 0 \\ * & * & 0 \\ * & * & * \end{pmatrix}, \qquad B = \begin{pmatrix} 1 & 0 & 0 \\ * & 1 & 0 \\ * & * & 1 \end{pmatrix}.$$

These groups have the following properties.

4.1

(a) $D(n, F)$ is abelian.
(b) $Tr_1(n, F)$ is nilpotent of class at most (actually equal to) $n - 1$. It is torsion-free if char $F = 0$ and it is a p-group of exponent p^m if char $F = p > 0$, where $p^{m-1} < n \leq p^m$.

B.A.F. Wehrfritz, *Group and Ring Theoretic Properties of Polycyclic Groups*, Algebra and Applications 10, DOI 10.1007/978-1-84882-941-1_4, © Springer-Verlag London Limited 2009

(c) $Tr_1(n, F) \triangleleft Tr(n, F)$ and $Tr(n, F) = D(n, F)[Tr_1(n, F)$. In particular $Tr(n, F)$
is soluble and $Tr(n, F)' \leq Tr_1(n, F)$. (In fact $Tr(n, F)' = Tr_1(n, F)$, unless
$F = GF(2)$ when $Tr(n, F) = Tr_1(n, F)$, and the derived length of $Tr(n, F)$ is
$1 - [-\log_2 n]$, unless $F = GF(2)$ when it is one less. Throughout $[r]$ for a real
number r denotes the greatest integer n with $n \leq r$.)

Proof Part (a) is trivial. Also the diagonal entries of the product of two lower triangular matrices are the product of the corresponding diagonal entries of the two
factors. Set $T = Tr(n, F)$ and $T_1 = Tr_1(n, F)$. Then for $a, b \in T$ we have $[a, b] \in T_1$
and so $T' \leq T_1$. In particular T_1 is normal in T. Clearly $T_1 \cap D = \langle 1 \rangle$ and $T = DT_1$.
Thus we have only to check that T_1 satisfies (b).

Let $e_1, e_2, \ldots, e_n$ be the standard basis of the n-row vector space $F^{(n)}$. Notice
that T_1 stabilizes the series

$$\{0\} < Fe_1 < Fe_1 \oplus Fe_2 < \cdots < (Fe_1 \oplus \cdots \oplus Fe_{n-1}) < F^{(n)}.$$

Then T_1 is nilpotent of class at most $n - 1$ by 1.19, is torsion-free if char $F = 0$ by
1.21(a) and is a p-group of finite exponent dividing p^{n-1} by 1.21(b) if char $F =
p > 0$.

On can also see all this and more by some direct calculations. Suppose a and b are
elements of T_1 with respectively $r - 1$ and $s - 1$ off diagonals of zeros immediately
below the main diagonal, so for example if $a = (a_{ij})$, then $a_{ij} = 0$ for $0 < i - j < r$.
If $a^{-1} = (a'_{ij})$, then $a'_{ij} = 0$ for $0 < i - j < r$ and $a'_{ij} = -a_{ij}$ for $i - j = r$. Similar
statements apply to b and s. Then direct calculation shows that if $c = [a, b] = (c_{ij})$,
then $c_{ij} = 1$ if $i = j$ and $c_{ij} = 0$ if $i < j$ or if $0 < i - j < r + s$. We obtain

$$\gamma^k T_1 \leq \{(a_{ij}) \in T_1 : a_{ij} = 0 \text{ for } 0 < i - j < k\}.$$

Hence $\gamma^n T_1 = \langle 1 \rangle$ and so T_1 is nilpotent of class at most $n - 1$. A little more work,
using $[1 + xe_{i,j}, 1 + e_{j,k}] = 1 + xe_{i,k}$ for $i > j > k$ and $x \in F^*$, shows that

$$\gamma^k T_1 = \{(a_{ij}) \in T_1 : a_{ij} = 0 \text{ for } 0 < i - j < k\},$$

so the class is exactly $n - 1$; $\{e_{i,j}\}$ is the set of standard matrix units. In the same
way

$$T_1^{(k)} = \{(a_{ij}) : a_{ij} = 0 \text{ for } 0 < i - j < \min\{n, 2^k\}\},$$

so T_1 has derived length d, where $2^{d-1} < n \leq 2^d$.

Let $x \in T_1$. By the Jordan Normal Form Theorem there exists an n by n invertible
matrix y over the algebraic closure of F such that

$$x^y = \text{diag}(J(r_1), J(r_2), \ldots, J(r_s)),$$

where $\sum_i r_i = n$ and $J(r)$ denotes the r by r Jordan matrix with eigenvalue 1 (so
$J(r) = (a_{ij})$ with $a_{ij} = 1$ if $0 \leq i - j \leq 1$ and $a_{ij} = 0$ otherwise). The entries
of $J(r)^m$ are given by binomial coefficients, so, for example its $(2, 1)$ entry is m,

its $(3, 1)$ entry is $m(m - 1)/2$ etc. Thus $J(r)$ has infinite order if char $F = 0$ and order p^t if char $F = p > 0$ where $p^{t-1} < n \leq p^t$. It follows that T_1 is torsion-free if char $F = 0$ and is a p-group of exponent p^m for $p^{m-1} < n \leq p^m$ if char $F = p > 0$. $\qquad \square$

Our discussion of the triangular groups in 4.1 is complete. Our objective now is to prove that any soluble linear group does not differ too greatly from a triangular group. Later we prove that any polycyclic group is isomorphic to a linear group, so we will be able to apply this to polycyclic groups.

4.2 **Theorem** (Lie-Kolchin, Mal'cev) Let G be a soluble subgroup of $GL(n, F)$ and let $F^\wedge$ denote an algebraic closure of the field F. Then there exists a normal subgroup G_0 of G of finite index and an element x in $GL(n, F^\wedge)$ such that $G_0^x \leq Tr(n, F^\wedge)$.

4.1 and 4.2 immediately yield the following.

4.3 **Corollary** If G is a soluble linear group, then $G \in \mathbf{NAF}$.

Kolchin (1948), extending work of Lie, proved that there is a particular canonical subgroup of G of finite index (its connected component of the identity relative to the Zariski topology and usually denoted by G^0, see appendix to this chapter), with the property of G_0 in 4.2. Its index in G is not boundable in terms of n only (if G is finite G^0 is always $\langle 1 \rangle$). Independently Mal'cev (1951), proved that there is an integer-valued function $\mu(n)$ of n only such that there is a subgroup G_0 as in 4.2 with $(G : G_0) \leq \mu(n)$. For proofs see Wehrfritz (1973a), 6.8 for the Lie-Kolchin version and 4.6 for the Mal'cev version. In the latter case it is not possible to choose G_0 canonically, except in special cases, see Wehrfritz (1978b). We do not need here the full strength of either of these two results and our particular choice of G_0 enables us to give a somewhat simpler proof. Our G_0 constructed below, although sort of canonical, will not usually be equal to G^0.

Let $V = F^{(n)}$ be the space of n-row vectors over F and suppose G is any subgroup of $GL(n, F)$. The elements of G act on V by right multiplication and V becomes a right G-module. Also G commutes with scalar multiplication, so in fact V becomes an FG-module (equivalently the given embedding $G \rightarrow GL(n, F)$ extends to a ring homomorphism of FG into the matrix ring $F^{n \times n}$).

Suppose $G^x \leq Tr(n, F)$ for some x in $GL(n, F)$. Now if $e_1, e_2, \ldots, e_n$ is the standard basis of V, set $E_i = F e_1 \oplus F e_2 \oplus \cdots \oplus F e_i$. Then

$$\{0\} = E_0 < E_1 < E_2 < \cdots < E_n = V$$

is a series of $Tr(n, F)$-invariant subspaces of V with one-dimensional factors. Set $V_i = E_i x^{-1}$. Then

$$\{0\} = V_0 < V_1 < V_2 < \cdots < V_n = V$$

is a series of FG-submodules of V with one-dimensional factors. Conversely, if we are given such a series we can pick v_i such that $V_i = Fv_i \oplus V_{i-1}$ for each i and then $v_1, v_2, \ldots, v_n$ is a basis of V. Thus there exists some x in $GL(n, F)$ with $v_i x = e_i$ for each i. Then $E_i x^{-1} G x \leq E_i$ for each i and so $G^x \leq Tr(n, F)$.

We use these remarks in the proof of 4.2. We also need the following.

4.3.1   Schur's lemma Let G be a subgroup of $GL(n, F)$, where F is an algebraically closed field, such that n-row the space $V = F^{(n)}$ over F is irreducible as FG-module. Then the centre $\zeta_1(G)$ of G lies in the group $F^* 1_n$ of scalar matrices.

Proof Let D denote the centralizer of G in the full matrix ring $F^{n \times n}$. If $d \in D \backslash \{0\}$, then d induces an FG-homomorphism of V, so $\ker d$ and $\operatorname{Im} d$ are FG-submodules of V. Hence $\ker d$ and $\operatorname{Im} d$ are $\{0\}$ or V. But $d \neq 0$. Thus $\ker d = \{0\}$, $\operatorname{Im} d = V$ and d is an FG-isomorphism. That is, D is a division ring.

Clearly $D \geq F 1_n \cong F$. Thus $F 1_n(d)$ is an extension field of F in D. It is also finite dimensional over F (clearly $F^{n \times n}$ has dimension n^2 over F). But F is algebraically closed, so there are no non-trivial finite field extensions of F. Therefore $F 1_n(d) = F 1_n$ and we have proved that $D = F 1_n$. Clearly $\zeta_1(G) \subseteq D$.   □

Proof of 4.2 Since $G \leq GL(n, F) \leq GL(n, F^\wedge)$, we may as well assume that F is algebraically closed. Let G_0 denote the intersection of all subgroups K of G of finite index that are the intersection with G of some F-subalgebra (containing 1_n) of $F^{n \times n}$. (There do exist such K, for example $K = G = G \cap F^{n \times n}$.) Any chain of distinct subspaces has length at most n^2. Hence the index $(G : G_0)$ is finite. Also G_0 is normal in G; in fact if N is a normal subgroup of G, then N_0 is a normal in G. Clearly $(G_0)_0 = G_0$.

How can we construct such subgroups K? Let Y be a subset of $F^{n \times n}$. Then the centralizer $C = \{x \in F^{n \times n} : xy = yx \text{ for all } y \in Y\}$ of Y in $F^{n \times n}$ is an F-subalgebra and $G \cap C = C_G(Y)$. Thus centralizers in G of finite index are one such possibility. Suppose $V \supseteq W \supset U$ are subspaces of V. Then $R = \{x \in F^{n \times n} : Ux \leq U \text{ and } Wx \leq W\}$ is an F-subalgebra of $F^{n \times n}$ and we have a natural map π of R into $\operatorname{End}_F(W/U)$. The inverse image under π of an F-subalgebra of $\operatorname{End}_F(W/U)$ is an F-subalgebra of $F^{n \times n}$. Thus, for example, if U and W are FG-submodules of V and if $\pi : G \to \operatorname{End}_F(W/U)$ is the natural map, then $(G_0)\pi \leq (G\pi)_0$.

Suppose first that V is FG_0-irreducible. Set $H = ((G_0)')_0$. Then H is normal in G_0. Let U be a minimal FH-submodule of V. Then U is FH-irreducible and so by induction on the derived length of G, $\dim_F U = 1$. If $g \in G_0$ then Ug is also a one-dimensional FH-submodule of V and $\sum Ug$ is an FG_0-submodule of V and hence is V itself. Therefore $V = U_1 \oplus U_2 \oplus \cdots \oplus U_n$ for some 1-dimensional FH-submodules U_i of V. Replacing G by a conjugate of itself (that is making a change of basis of V) we may assume that $U_i = Fe_i$ for each i, $\{e_i : 1 \leq i \leq n\}$ being as usual the standard basis of V. Then $H \leq D(n, F)$. If $h \in H$ and $g \in G_0$ then h and h^g have the same eigenvalues to the same multiplicities. Thus h^g has the same entries as h, but possibly in a different order. Therefore the centralizer of h in G_0 has finite index in G_0 and consequently contains $(G_0)_0 = G_0$. Thus $H \leq \zeta_1(G_0)$. By Schur's Lemma $H \subseteq F 1_n$.

Let $h = \eta 1_n \in H$. Now $H \leq (G_0)'$, so $\det h = 1$ (compute any $\det[x, y]$). That is $\eta^n = 1$ and H is finite. Consequently $(G_0)'$ is finite, its centralizer has finite index in G_0 and $(G_0)' \leq \zeta_1(G_0)$. But now for any x in G_0 the map $g \mapsto [g, x]$ is a homomorphism (by 1.2) of G_0 and its kernel (the centralizer of x) has finite index in G_0 and so contains $(G_0)_0 = G_0$. Consequently $G_0 = \zeta_1(G_0)$, which is scalar by 4.3.1 again. Since V is FG_0-irreducible, we have $\dim_F V = 1$.

We now return to the general case. Clearly there exists a series

$$\{0\} = V_0 < V_1 < V_2 < \cdots < V_r = V$$

of FG_0-submodules such that each V_i/V_{i-1} is FG_0-irreducible. If π_i denotes the natural map of G_0 into $\mathrm{End}_F(V_i/V_{i-1})$ then by the above $G_0\pi_i = (G_0)_0\pi_i \leq (G_0\pi_i)_0$. Thus $G_0\pi_i = (G_0\pi_i)_0$ and the above yields that each $\dim_F(V_i/V_{i-1}) = 1$. The theorem follows from our remarks before the statement of 4.3.1.   □

4.4   Theorem (Mal'cev 1951) Every soluble subgroup of $GL(n, \mathbf{Z})$ is polycyclic.

Proof Let $G \leq GL(n, \mathbf{Z})$ be soluble. It suffices to prove that G is finitely generated, for then every subgroup of G is finitely generated in the same way and so

$$G \in \mathbf{S} \cap \mathbf{Max} \subseteq P(\mathbf{G} \cap \mathbf{A}) = \mathbf{P}.$$

Let A denote an algebraic closure of $\mathbf{Q}$. By 4.2 there is a normal subgroup H of G of finite index and an element x of $GL(n, A)$ such that $H^x \leq Tr(n, A)$. It suffices to prove that H is finitely generated. Let F be the subfield generated by the $2n^2$ entries in x and x^{-1} and let J be the ring of algebraic integers of F (that is, the set of elements of F that are roots of monic polynomials with integer coefficients). The Dirichlet Unit Theorem (see Samuel (1972), p. 60 or any book on algebraic number theory) states that the group $U(J)$ of units of J is a finitely generated (abelian) group.

Now $H \leq GL(n, \mathbf{Z})$, so $H^x \leq GL(n, F) \cap Tr(n, A) = Tr(n, F)$. If $g \in H$, the eigenvalues of g are the diagonal entries of g^x and are also the roots of its characteristic polynomial $\det(X1_n - g) \in \mathbf{Z}[X]$. Thus the diagonal entries of g^x lie in J. This also applies to $(g^{-1})^x$, so in fact they lie in $U(J)$. Thus H^x modulo $Tr_1(n, F)$ embeds into the direct product of n copies of $U(J)$. Set $U = H \cap xTr_1(n, F)x^{-1}$. Then U is normal in H and H/U is a finitely generated abelian group.

With $e_1, e_2, \ldots, e_n$ denoting the standard basis as usual, set $E_i = Fe_1 \oplus \cdots \oplus Fe_i$ and $V_i = E_ix^{-1} \cap \mathbf{Z}^{(n)}$. Then $Tr_1(n, F)$ stabilizes the series $\{0\} < E_1 < \cdots < E_n = V$. Now $U \leq H \leq GL(n, \mathbf{Z})$ and $U^x \leq Tr_1(n, F)$. Therefore U stabilizes the series

$$\{0\} < V_1 < V_2 < \cdots < V_n = V.$$

The following lemma completes the proof.   □

4.4.1 Let $V = V_0 \geq V_1 \geq \cdots \geq V_n = V$ be a series of normal subgroups of the group V with $V \in \mathbf{Max}$. If U is a subgroup of Aut V stabilizing this series, then U is finitely generated.

Proof If $n \leq 1$ then $U = \langle 1 \rangle$. By induction on n we may assume that $U/C_U(V_1)$ is finitely generated. Suppose V can be generated by r elements. Now $C_U(V_1)$ stabilizes the series $\langle 1 \rangle \leq V_1 \leq V$, so by 1.20(c) the group $C_U(V_1)$ embeds into the direct product of r copies of the finitely generated abelian group $\zeta_1(V_1)$. Therefore $C_U(V_1)$ is finitely generated, so U is finitely generated and the proof is complete. $\square$

Our next main objective is to prove the converse of 4.4; namely that each polycyclic group is isomorphic to a soluble subgroup of some $GL(n, \mathbf{Z})$. Once we have done this it will follow from 4.3 that $\mathbf{P} \subseteq \mathbf{NAF}$; that is, that polycyclic groups are nilpotent-by-abelian-by-finite. Unfortunately we have to prove this containment first. Below $\mathbf{C}$ denotes the complex numbers.

4.5 (Mal'cev 1951) $\mathbf{PF} \subseteq \mathbf{NAF}$.

Proof Clearly we may just consider a polycyclic group G. Now G contains by 2.6 a series $\langle 1 \rangle = K_0 \leq K_1 \leq \cdots \leq K_t \leq G$ of normal subgroups with G/K_t finite and each K_i/K_{i-1} free abelian of finite rank, n_i say. Then $G/C_G(K_i/K_{i-1})$ embeds into $\mathrm{Aut}(K_i/K_{i-1}) \cong GL(n_i, \mathbf{Z}) \leq GL(n_i, \mathbf{C})$. By 4.2 there is a triangularizable (over $\mathbf{C}$) normal subgroup of its image in $GL(n_i, \mathbf{C})$. Let T_i be its inverse image in G. Then T_i is normal of finite index in G and $(T_i)'$ stabilizes a series in n_i-row space over $\mathbf{C}$ and hence over $\mathbf{Z}$ and hence in K_i/K_{i-1} of length n_i. Set $T = T_1 \cap T_2 \cap \cdots \cap T_t \cap K_t$. Then T is a normal subgroup of G of finite index and T' stabilizes a series running from $\langle 1 \rangle$ to G (of length $1 + \sum_i n_i$). Thus T' is nilpotent by 1.19 and $G \in \mathbf{NAF}$. $\square$

4.6 Let H be a subgroup of the group G of finite index m such that H is isomorphic to a subgroup of $GL(n, \mathbf{Z})$. Then G is isomorphic to a subgroup of $GL(mn, \mathbf{Z})$.

Here the integers $\mathbf{Z}$ could be replaced by any ring.

Proof Let $\phi : H \to GL(n, \mathbf{Z})$ be the given embedding. Then via ϕ the n-row space $\mathbf{Z}^{(n)}$ becomes a faithful H-module and $\mathbf{Z}^{(n)} \otimes_{\mathbf{Z}H} \mathbf{Z}G$ becomes a faithful G-module that is a free $\mathbf{Z}$-module of rank mn. Alternatively let $g_1, g_2, \ldots, g_m$ be a right transversal of H to G. If $g \in G$, then $g_i g = h_i g_{i\sigma}$ for some $h_i \in H$ and $\sigma \in \mathrm{Sym}(m)$. Define $\psi : G \to GL(mn, \mathbf{Z})$ by $g\psi = (g(i, j))_{i, j = 1, 2, \ldots, m}$ where $g(i, j)$ is $h_i\phi$ if $j = i\sigma$ and is the n by n zero matrix 0_n otherwise. $\square$

Exercise Check that ψ is an embedding.

4.7 Let $G = \langle g_1; g_2, \ldots, g_n \rangle$ be a finitely generated group and $\mathbf{a}$ an ideal of the group ring $\mathbf{Z}G$ with $\mathbf{Z}G/\mathbf{a}$ finitely generated as $\mathbf{Z}$-module. Then $\mathbf{a}$ is finitely generated as an ideal and $\mathbf{Z}G/\mathbf{a}^r$ is finitely generated as a $\mathbf{Z}$-module for each positive integer r.

Proof By hypothesis $\mathbf{Z}G = u_1\mathbf{Z} + u_2\mathbf{Z} + \cdots + u_m\mathbf{Z} + \mathbf{a}$ for some $u_1, u_2, \ldots, u_m$. Set

$$M = u_1\mathbf{Z} + u_2\mathbf{Z} + \cdots + u_m\mathbf{Z} + g_1\mathbf{Z} + g_2\mathbf{Z} + \cdots + g_n\mathbf{Z} + g_1^{-1}\mathbf{Z} + g_2^{-1}\mathbf{Z} + \cdots + g_n^{-1}\mathbf{Z}.$$

Certainly $\mathbf{Z}G = M + \mathbf{a}$. If $m_1, m_2 \in M$, then $m_1 m_2 = x + a$, for some $x \in M$ and $a \in \mathbf{a}$. Let $M^2 = \sum_{m, m' \in M} mm'\mathbf{Z}$. Then $a = m_1 m_2 - x \in \mathbf{a} \cap (M + M^2)$ and the latter is a finitely generated $\mathbf{Z}$-module (since $M + M^2$ is). Let $\mathbf{b}$ be the ideal of $\mathbf{Z}G$ generated by $\mathbf{a} \cap (M + M^2)$. Then $\mathbf{b}$ is a finitely generated ideal and $(M + \mathbf{b})(M + \mathbf{b}) \subseteq M + \mathbf{b}$ by construction of $\mathbf{b}$. Also all g_i and g_i^{-1} lie in M, so $M + \mathbf{b} = \mathbf{Z}G$. Therefore $\mathbf{Z}G/\mathbf{b}$ is finitely generated over $\mathbf{Z}$. Trivially $\mathbf{b} \leq \mathbf{a}$, so $\mathbf{a}/\mathbf{b}$ is finitely $\mathbf{Z}$-generated and hence $\mathbf{a}$ is a finitely generated ideal of $\mathbf{Z}G$.

We are given that $\mathbf{Z}G/\mathbf{a}$ is finitely $\mathbf{Z}$-generated. Suppose $\mathbf{Z}G/\mathbf{a}^r$ for some $r \geq 1$ is also finitely $\mathbf{Z}$-generated, say $\mathbf{Z}G = v_1\mathbf{Z} + v_2\mathbf{Z} + \cdots + v_s\mathbf{Z} + \mathbf{a}^r$. Suppose also that $w_1, w_2, \ldots, w_t$ generate $\mathbf{a}$ as an ideal. Then

$$\mathbf{a} = \mathbf{Z}G(w_1\mathbf{Z} + w_2\mathbf{Z} + \cdots + w_t\mathbf{Z})\mathbf{Z}G = \left(\sum_{i,j,k} v_i w_j v_k \mathbf{Z} \right) + \mathbf{a}^{r+1}.$$

Thus $\mathbf{a}/\mathbf{a}^{r+1}$ is finitely $\mathbf{Z}$-generated and consequently so too is $\mathbf{Z}G/\mathbf{a}^{r+1}$. Induction on r completes the proof. □

4.8 Theorem (Auslander 1967) If G is a polycyclic-by-finite group, there exists an integer n such that G is isomorphic to a subgroup of $GL(n, \mathbf{Z})$.

Thus this is our promised converse of 4.4. Almost immediately Swan (1967) produced a considerably more elementary proof. It had been an open question for some 16 years. I believe it was first conjectured by P. Hall. Anyway Hall (1957) proved it for $\mathbf{G} \cap \mathbf{NF}$ groups. Other special cases were proved in Learner (1962). Earlier Wang (1956) had proved that $\mathbf{PF}$ groups could at least be embedded into some $GL(n, \mathbf{C})$, $\mathbf{C}$ being the complex numbers,

Proof The proof we present here is based on Swan's. Suppose we can prove the following.

4.8.1 Let N and H be normal subgroups of a group K with $K' \leq N \leq H$ and K/H infinite cyclic. Suppose H is a subgroup of $GL(n, \mathbf{Z})$ with N unipotent (meaning that N stabilizes a series in $\mathbf{Z}^{(n)}$). Then there exists an integer m and an embedding ϕ of K into $GL(m, \mathbf{Z})$ such that $N\phi$ is unipotent. Moreover if $H = N$ and K is nilpotent then we can choose ϕ so that $K\phi$ itself is unipotent.

Note that N is nilpotent by 1.19 and H and K are polycyclic by 4.4. If G is as in 4.8, then by 2.6 and 4.5 there is a series

$$\langle 1 \rangle = G_0 < G_1 < \cdots < G_r < \cdots < G_s \leq G$$

of subgroups of G with G/G_s finite, each G_i/G_{i-1} infinite cyclic, G_r nilpotent and $G_s' \leq G_r$. Suppose we have an embedding of G_{i-1} into $GL(n, \mathbf{Z})$ with $G_{i-1} \cap G_r$ unipotent. Set $H = G_{i-1}$, $N = G_{i-1} \cap G_r$ and $K = G_i$. By 4.8.1 there exists m and an embedding τ of G_i into $GL(m, \mathbf{Z})$. If $i > r$ then $G_{i-1} \cap G_r = G_r = N$ and $N\tau$ is chosen unipotent. If $i \leq r$, then $H = N$ and K is nilpotent. Thus we may choose τ so that $G_i \tau$ is unipotent. Trivially we have an embedding of G_0 into $GL(1, \mathbf{Z})$ with its image unipotent. Hence by induction there is an embedding of G_s into $GL(t, \mathbf{Z})$ for some t. Then by 4.6 the group G is isomorphic to a subgroup of $GL(t(G : G_s), \mathbf{Z})$. The proof will be complete.

Proof of 4.8.1 Now as above H is polycyclic by 4.4. The inclusion $H \leq GL(n, \mathbf{Z})$ determines a ring homomorphism θ of $\mathbf{Z}H$ into the matrix ring $\mathbf{Z}^{n \times n}$. Set $\mathbf{k} = \ker \theta$. Let $\mathbf{n}$ be the ideal of $\mathbf{Z}H$ generated by all $x - 1$ for x in N. Now N stabilizes a series $\{0\} = V_0 < V_1 < \cdots < V_{n'} = \mathbf{Z}^{(n)}$ in $\mathbf{Z}^{(n)}$.

Exercise One can always choose $n' = n$.

Then if $x \in N$ we have $V_i(x - 1) \leq V_{i-1}$ for each $i > 0$. Also if $h \in H$, then $h(x^h - 1) = (x - 1)h$ and $x^h \in N$ since $N \triangleleft H$. Therefore $\mathbf{Z}^{(n)} \mathbf{n}^{n'} = \{0\}$; that is, $\mathbf{n}^{n'} \leq \mathbf{k}$.

$\mathbf{Z}H/(\mathbf{n} + \mathbf{k})$ is an image of $\mathbf{Z}H/\mathbf{k} \cong \mathrm{Im}\,\theta \leq \mathbf{Z}^{n \times n}$ and so is finitely $\mathbf{Z}$-generated. Hence by 4.7 the factor $\mathbf{Z}H/(\mathbf{n} + \mathbf{k})^{n'}$ is finitely $\mathbf{Z}$-generated. Let $T/(\mathbf{n} + \mathbf{k})^{n'}$ be its torsion subgroup. Then $V = \mathbf{Z}H/T$ is free abelian of finite rank. Since $(\mathbf{n} + \mathbf{k})^{n'} \leq \mathbf{k}$ and $\mathbf{Z}H/\mathbf{k}$ embeds into $\mathbf{Z}^{n \times n}$ and hence is $\mathbf{Z}$-torsion-free, so $T \leq \mathbf{k}$ and V is faithful as H-module. Also N stabilizes a series in V, namely $\{V\mathbf{n}^i : 0 \leq i \leq n'\}$, since $\mathbf{n}^{n'} \leq T$.

We have now to make V into a K-module. K/H is infinite cyclic, so $K = \langle g \rangle [H$ for some g of infinite order. Make $\mathbf{Z}H$ into a right K-module by letting g act by conjugation and H by right multiplication. Explicitly if $k \in K$ then $k = g^j h$ for some unique $j \in \mathbf{Z}$ and $h \in H$. Set $(\sum_i \alpha_i h_i).k = \sum_i \alpha_i (g^{-j} h_i g^j)h$. Then $\mathbf{Z}H$ becomes a right K-module (check). For $h \in H$,

$$h.g = g^{-1}hg = h[h, g] = h + h([h, g] - 1) \in h + \mathbf{n},$$

since $K' \leq N$. Thus g fixes every element of $\mathbf{Z}H/\mathbf{n}$ and so $\mathbf{n} + \mathbf{k}$ is a K-submodule. It follows that $(\mathbf{n} + \mathbf{k})^{n'}$ is a K-submodule and hence that T is too. Therefore $V = \mathbf{Z}H/T$ is a K-module.

Unfortunately it may not be a faithful K-module. Let $W = \mathbf{Z} \oplus \mathbf{Z}$ and make W into a K-module by letting H act trivially and g act as the matrix $\begin{pmatrix} 1 & 0 \\ 1 & 1 \end{pmatrix}$. Note that $\begin{pmatrix} 1 & 0 \\ 1 & 1 \end{pmatrix}^p = \begin{pmatrix} 1 & 0 \\ p & 1 \end{pmatrix}$ for all p in $\mathbf{Z}$, so W is a faithful $\langle g \rangle$-module. Then $V \oplus W$ is a K-module. If $k \in K$ acts trivially on $V \oplus W$, then $k \in H$ since W is K/H-faithful, and V is H-faithful. Thus $k = 1$ and $V \oplus W$ is K-faithful. N acts unipotently on V and trivially on W, so N acts unipotently on $V \oplus W$. Set $m = \mathrm{rank}(V \oplus W)$. Then $\mathrm{Aut}(V \oplus W)$ is isomorphic to $GL(m, \mathbf{Z})$. Consequently we have constructed a suitable embedding ϕ of K into $GL(m, \mathbf{Z})$.

We are left with the case where $H = N$ and K is nilpotent. We have to prove that K acts unipotently on $V \oplus W$. It suffices to prove that $X = K[(V \oplus W)$ is nilpotent. Set $Y = H[(V \oplus W)$. Then Y is a normal subgroup of X in the obvious way. Also $H = N$ is nilpotent and acts unipotently on $V \oplus W$, so Y is nilpotent. Further

$$Y' \geq [V \oplus W, N] \geq V\mathbf{n} \oplus W\mathbf{n} = V\mathbf{n}.$$

Now $H = N$ acts trivially on $V/V\mathbf{n}$ and so does g. Hence K centralizes $V/V\mathbf{n}$; it also acts unipotently on W. Thus $X/V\mathbf{n}$ is nilpotent and consequently so too is X/Y'. Then X is nilpotent by 1.23. The proof of 4.8 is complete. $\qquad\square$

It is possible to avoid using the theorem 1.23 of P. Hall by means of a direct, elementary but unpleasant calculation, see 2.7 of Wehrfritz (1974) and Point 2 of the proof of 2.5 in Wehrfritz (1973a).

The notion of unipotence is more important than the passing reference in 4.8.1 and the other occasional references in this book might imply. In general for any field F, a subgroup of $GL(n, F)$ is *unipotent* if it stabilizes a series of subspaces in the row vector space $V = F^{(n)}$. Thus $U \leq GL(n, F)$ is unipotent if and only if $U^x \leq Tr_1(n, F)$ for some x in $GL(n, F)$, that is, if U is unitriangularizable over F. Easy but not quite trivial (e.g. see Wehrfritz 1973a, pp. 13 and 14) is that this is equivalent to $(u - 1)^n = 0$ for every $u \in U$. A fourth equivalent definition is that $V\mathbf{u}^n = \{0\}$ for $\mathbf{u}$ the augmentation ideal of U in FU. Any subgroup of $G \leq GL(n, F)$ normalizing U clearly normalizes $\mathbf{u}$ and hence G acts on each $V\mathbf{u}^{i-1}/V\mathbf{u}^i$. It follows easily that G has a unique maximal unipotent normal subgroup $u(G)$, its *unipotent radical*. By Clifford's theorem $G/u(G)$ is isomorphic to a completely reducible subgroup of $GL(n, F)$.

We now consider some simple applications of 4.8. The first is to give a second proof of 2.10. Let G be a polycyclic-by-finite group. We have to prove that G is residually finite. By 4.8 we can regard G as a subgroup of $GL(n, \mathbf{Z})$ for some n. If p is any prime we have a map π_p of $GL(n, \mathbf{Z})$ into $GL(n, p)$ obtained by reducing all matrix entries modulo p. Certainly $\mathrm{Im}\,\pi_p$ is finite, so $K_p = \ker \pi_p$ is a normal subgroup of $GL(n, \mathbf{Z})$ of finite index. Let $k = (k_{ij}) \in \bigcap_p K_p$. If $i \neq j$, then p divides k_{ij} for all primes p, so $k_{ij} = 0$. Also $(k_{ii} - 1)$ is divisible by p for each i and all primes p, so each $k_{ii} = 1$. Therefore $k = 1$, $\bigcap_p K_p = \langle 1 \rangle$, $GL(n, \mathbf{Z})$ is residually finite and G is residually finite. In particular it is worth recording the following.

4.9 $GL(n, \mathbf{Z})$ is residually finite.

4.10   For any prime p and positive integer n we have $GL(n, \mathbf{Z}) \in (RF_{\{p\}})\mathbf{F}$.

2.11 follows at once from 4.8 and 4.10 (2.11 says $\mathbf{PF} \subseteq \bigcap_p (RF_{\{p\}})\mathbf{F}$).

Proof As above we have $\pi_p : GL(n, \mathbf{Z}) \to GL(n, p)$. We prove that $K = \ker \pi_p$ is residually a finite p-group. Now $K \subseteq 1_n + p\mathbf{Z}^{n \times n}$. Thus K stabilizes the series

$$\mathbf{Z}^{(n)} \geq p\mathbf{Z}^{(n)} \geq p^2\mathbf{Z}^{(n)} \geq \cdots \geq p^i\mathbf{Z}^{(n)} \geq \cdots.$$

Then $K/C_K(\mathbf{Z}^{(n)}/p^i\mathbf{Z}^{(n)})$ is a p-group by 1.21, necessarily finite since $\mathbf{Z}^{(n)}/p^i\mathbf{Z}^{(n)}$ is finite. If $k \in \bigcap_i C_K(\mathbf{Z}^{(n)}/p^i\mathbf{Z}^{(n)})$, then $[\mathbf{Z}^{(n)}, k] \leq \bigcap_i p^i\mathbf{Z}^{(n)} = \{0\}$. Thus $k = 1$ and the result follows.   □

There have been a number of developments of 4.8. Merzljakov (1970) proved that if G is a polycyclic-by-finite group, then Aut G, or more generally the holomorph Hol $G = ($Aut $G)[G$, is isomorphic to a subgroup of some $GL(n, \mathbf{Z})$. (Learner (1962) had proved this earlier for G finitely generated and nilpotent and Merzljakov himself for G supersoluble in 1968.) This theorem of Merzljakov, together with results of A. Borel and Harish-Chandra, can be used to give an alternative proof of the following theorem of Auslander (1969). If G is a polycyclic-by-finite group, then Aut G is finitely presented (and in particular is finitely generated), see Wehrfritz (1973b). Similar results hold for related classes of groups. For example (see Wehrfritz 1974); if G is a finite extension of a torsion-free soluble group of finite rank, then Hol G embeds in $GL(n, \mathbf{Q})$ for some n. The same paper and Wehrfritz (1973b) contain alternative proofs of Merzljakov's Theorem. Also if G is a polycyclic-by-finite group, its outer automorphism group Out G embeds into some $GL(n, \mathbf{Z})$, see Wehrfritz (1994). This is not an immediate consequence of Merzljakov's theorem and does not extend to torsion-free soluble groups of finite rank, for example. If G is a Chernikov group (that is, a soluble-by-finite group with the minimal condition on subgroups) there exists n such that G and Out G both embed into $GL(n, \mathbf{C})$, while Aut G is not in general isomorphic to any linear group, see Kegel and Wehrfritz (1973), p. 109.

Appendix on the Zariski Topology

We have avoided using this so far and will continue to do so except incidentally and very briefly much later in this book. It does lead, however, to very useful language and techniques. This current chapter on linear groups is the most appropriate place to introduce it. I anticipate most readers will probably wish to skip this section until they actually need it.

Let F be a field, m a positive integer, $U = F^{(m)}$, the space of m-row vectors over F, and $R = F[X_1, X_2, \ldots, X_m]$, the polynomial ring over F in the X_i. Say $A \subseteq U$ is closed (in U) if there exists $S \subseteq R$ such that

$$A = \{(a_1, a_2, \ldots, a_m) \in U : f(a_1, a_2, \ldots, a_m) = 0 \text{ for all } f \text{ in } S\} = V(S), \text{ say.}$$

This puts a T_1-topology (a topology with 1-element sets closed) on U; for $\bigcap_\alpha V(S_\alpha) = V(\bigcup_\alpha S_\alpha)$ is closed, as is $V(S) \cup V(T) = V(ST)$ and clearly $U = V(0)$ and $\emptyset = V(1)$. Also $V(X_1 - a_1, X_2 - a_2, \ldots, X_m - a_m) = \{(a_1, a_2, \ldots, a_m)\}$

so the topology is T_1 (but not Hausdorff). Now $V(S) = V$(ideal of R generated by S); in fact $\{g \in R : g(a) = 0 \text{ for all } a \in V(S)\}$ is an ideal of R. By the Hilbert Basis Theorem R satisfies the ascending chain condition on ideals, cf. 3.7.1. Therefore our topology satisfies the descending chain condition on closed sets (equivalently the ascending chain condition on open sets). This topology is called the Zariski topology on U.

Suppose in addition to the above notation that W is n-row space over F, also taken with its Zariski topology, and let $p_1/q_1, p_2/q_2, \ldots, p_n/q_n$ be n rational functions over F in our m indeterminates $X_1, X_2, \ldots, X_m$, the p_i and q_j lying in R. Suppose $S = V(q_1 q_2, \ldots, q_n)$ and set $T = U \backslash S$. Then

$$\phi : (a_1, a_2, \ldots, a_m) \mapsto ((p_1/q_1)(a_1, a_2, \ldots, a_m), \ldots, (p_n/q_n)(a_1, a_2, \ldots, a_m))$$

is a continuous map of T into W. (We remove S since ϕ will not be defined on S.)

For if B is a closed subset of W, then by the Hilbert Basis Theorem $B = V(g_1, g_2, \ldots, g_t)$ for some finitely many elements g_j of $F[Y_1, Y_2, \ldots, Y_n]$. Let k be the maximal total degree of any of the g_j. Then

$$h_j = q_1^k q_2^k \ldots q_n^k g_j(p_1/q_1, p_2/q_2, \ldots, p_n/q_n) \in F[X_1, X_2, \ldots, X_m]$$

and

$$B\phi^{-1} = \{x \in T : g_j(p_1/q_1, p_2/q_2, \ldots, p_n/q_n)(x) = 0 \text{ for } j = 1, 2, \ldots, t\}$$
$$= \{x \in T : h_j(x) = 0 \text{ for } j = 1, 2, \ldots, t\},$$

which is closed in T. The continuity of ϕ follows.

Exercise Prove that any subspace V of U is closed in U and that the Zariski topology on U induces on V the Zariski topology on V.

Let G be a subgroup of $GL(n, F)$. Then $F^{n \times n}$ is an n^2-dimensional space over F and as such carries its Zariski topology. This topology induces a topology on its subset G, which we call the Zariski topology on G. If $a \in G$ then the four maps

$$x \mapsto ax, \qquad x \mapsto x^{-1},$$
$$x \mapsto xa, \qquad x \mapsto x^{-1}ax$$

of G to itself are all continuous since these maps are given either by polynomials or by rational functions in which the denominator is $\det x$ and consequently is non-zero on G. In fact the first three maps are homeomorphisms of G.

Exercise Prove that a change of basis for $F^{(n)}$ or a ground field extension of F does not change the topology on G. Thus we can talk of the Zariski topology on G.

4.11 Let G be a subgroup of $GL(n, F)$ (taken with its Zariski topology). Let C be the connected component of G containing 1. Then C is a normal subgroup of G of finite index and the connected components of G are precisely the cosets of C.

The standard notation for this C is G^0.

Exercise If G is also finite prove that $G^0 = \langle 1 \rangle$.

Proof By the descending chain condition on closed sets, G has a minimal (non-empty) open closed subset D. Clearly such a D is connected. Again $G \backslash D$ also has the descending chain condition on closed sets. Thus we can repeatedly pick minimal open closed subsets $D_1, D_2, \ldots, D_r$ of G that are pairwise disjoint. By the ascending chain condition on open sets this process must halt and we have $G = D_1 \cup D_2 \cup \cdots \cup D_r$ for some r. Then the D_i are the connected components of G and in particular G has only finitely many of such.

A continuous image of a connected set is connected. Thus $c^{-1}C$ and $x^{-1}Cx$ are connected for any c in C and x in G. Also $1 \in c^{-1}C \cap C$, so $c^{-1}C \cup C$ is connected and so $c^{-1}C \subseteq C$. Therefore C is a subgroup of G. Further $1 \in C \cap x^{-1}Cx$, so $x^{-1}Cx \subseteq C$ and hence C is a normal subgroup of G.

Finally $x \mapsto ax$ is a homeomorphism of G for any $a \in G$, so aC is the connected component of G containing a. Now G is the disjoint union of both the cosets of C and the connected components of G. This proves that the cosets of C and the connected components of G are one and the same. In particular $(G : C) = r < \infty$. The proof is complete. □

4.12 Let H be a subgroup of the subgroup G of $GL(n, F)$. The following are equivalent.

(a) $G^0 \leq H$.
(b) H is closed in G and the index $(G : H)$ is finite.
(c) H is open in G.

Proof Suppose (a) holds. Certainly $(G : H) \leq (G : G^0)$ is finite. Also G^0 is closed, as are its cosets in G, and H is a union of finitely many of these cosets of G^0. Therefore H is closed. Thus (b) holds.

Suppose (b) holds. Now H and its cosets are closed in G. Also $G \backslash H$ is the union of all the finitely many cosets of H except H itself. Thus $G \backslash H$ is closed and hence H is open.

Suppose (c) holds. The cosets of H are now open and $G \backslash H$ is a union of (potentially infinitely many) cosets of H. Thus $G \backslash H$ is open and hence H is closed. But now $G^0 \cap H$ is non-empty (it contains 1) and is open and closed in G^0. Therefore it is G^0 and consequently $G^0 \leq H$. The proof is complete. □

4.13 Theorem (Lie-Kolchin) Let G be a soluble subgroup of $GL(n, F)$, where F is algebraically closed. Then there exists x in $GL(n, F)$ such that $x^{-1}G^0x \leq Tr(n, F)$.

Proof In 4.2 we constructed a subgroup G_0 of G of finite index and an x in $GL(n, F)$ with $x^{-1}G_0 x \leq Tr(n, F)$. Now G_0 is defined as the intersection of G with a subspace (indeed an F-subalgebra) of $F^{n \times n}$. Thus G_0 is closed in G and $G^0 \leq G_0$ by 4.12. The theorem follows. $\qquad\square$

An important aspect of the Zariski topology for us is that the properties of the closure of a subgroup closely mirror the properties of the original subgroup. We close this appendix by recording some of these properties.

4.14 Let G be a subgroup of $GL(n, F)$. The normalizer of a closed subset S of G in G is closed in G. The centralizer of any subset T of G in G is closed in G.

Proof Let $S(a)$ be the inverse image of S in G under the continuous map $x \mapsto x^{-1}ax$. Then $N_1 = \bigcap_{a \in S} S(a) = \{x \in G : x^{-1}Sx \subseteq S\}$ is closed in G. If $(a)S$ is the inverse image of S under $x \mapsto xax^{-1}$ (which is continuous note, being the composite of $x \mapsto x^{-1}$ and $x \mapsto x^{-1}ax$) then $N_2 = \bigcap_{a \in S}(a)S = \{x \in G : x^{-1}Sx \supseteq S\}$ is closed. Consequently $N_G(S) = N_1 \cap N_2$ is closed in G.

Finally $C_G(T) = \bigcap_{t \in T} C_G(t) = \bigcap_{t \in T} N_G(\{t\})$ and G is T_1. Therefore by the first part $C_G(T)$ is closed in G. $\qquad\square$

Exercise If $G \leq GL(n, F)$ is an FC-group (meaning that g^G is finite for all g in G), prove that G is centre by finite.

4.15 Let G be a subgroup of $GL(n, F)$, H a subgroup of G and $H^\wedge$ the closure of H in G. Then $H^\wedge$ is a subgroup of G. If H is normal in G, then so is $H^\wedge$.

Proof Now $x \mapsto x^{-1}$ is a homeomorphism of G and $H = H^{-1}$. Therefore $(H^\wedge)^{-1} = H^\wedge$. Let $h \in H$. Then $H \subseteq H^\wedge(x \mapsto hx)^{-1}$ (meaning the inverse image of $H^\wedge$ under the map $x \mapsto hx$) and the latter is closed. Thus $H^\wedge \subseteq H^\wedge(x \mapsto hx)^{-1}$. That is $hH^\wedge \subseteq H^\wedge$ and so $HH^\wedge \subseteq H^\wedge$. Let $k \in H^\wedge$. This says that $H \subseteq H^\wedge(x \mapsto xk)^{-1}$, so $H^\wedge \subseteq H^\wedge(x \mapsto xk)^{-1}$, $H^\wedge k \subseteq H^\wedge$, $H^\wedge H^\wedge \subseteq H^\wedge$ and $H^\wedge$ is a subgroup of G.

Suppose $H \triangleleft G$. If $g \in G$, then $H \subseteq H^\wedge(x \mapsto g^{-1}xg)^{-1}$ (notice that this map is continuous being again the composite of two of our standard maps), so $H^\wedge \subseteq H^\wedge(x \mapsto g^{-1}xg)^{-1}$. Consequently $g^{-1}H^\wedge g \leq H^\wedge$ for all g in G and therefore $H^\wedge$ is normal in G. $\qquad\square$

4.16 Let A, B, and C be subgroups of the subgroup G of $GL(n, F)$, with closures $A^\wedge$, $B^\wedge$ and $C^\wedge$ respectively in G. If $[A, C] \leq B$, then $[A^\wedge, C^\wedge] \leq B^\wedge$.

Note that 4.16 generalizes by 1.4 the normality part of 4.15

Proof Let $c \in C$. Then $A \subseteq B^\wedge(x \mapsto [x, c] = c^{-x}c)^{-1}$, so $A^\wedge \subseteq B^\wedge(x \mapsto [x, c])^{-1}$. That is, $[A^\wedge, C] \leq B^\wedge$. If $a \in A^\wedge$, this implies that $C \subseteq B^\wedge(x \mapsto [a, x] = a^{-1}a^x)^{-1}$, so $C^\wedge \subseteq B^\wedge(x \mapsto [a, x])^{-1}$. Hence $[A^\wedge, C^\wedge] \leq B^\wedge$. $\qquad\square$

4.17 Let G be a subgroup of $GL(n, F)$, H a subgroup of G and $K = H^{\wedge}$ the closure of H in G. Then;

(a) H is soluble of derived length d if and only if K is too, and
(b) H is nilpotent of class c if and only if K is too.

In particular H is abelian if and only if K is abelian.

Proof Suppose $\langle 1 \rangle = H_0 \vartriangleleft H_1 \vartriangleleft \cdots \vartriangleleft H_r = H$ is a series of H. Let K_i denote the closure of H_i in G. Now $[H_i, H_{i+1}] \leq H_i$, so by 4.16 we have $[K_i, K_{i+1}] \leq K_i$. Then $K_i \vartriangleleft K_{i+1}$, see 1.4. Also $K_0 = \langle 1 \rangle$ since the topology is T_1. Consequently $\langle 1 \rangle = K_0 \vartriangleleft K_1 \vartriangleleft \cdots \vartriangleleft K_r = K$ is a series for K.

(a) If each H_i / H_{i-1} is abelian, then $[H_i, H_i] \leq H_{i-1}$, so by 4.16 we have $[K_i, K_i] \leq K_{i-1}$. Thus if H is soluble of derived length d, then K is soluble of derived length at most d. But $H \leq K$, so K will have derived length exactly d. If K is soluble of derived length d, then, as $H \leq K$, so H is soluble of derived length some $d' \leq d$. But then by the first part K also has derived length d'. Therefore $d' = d$.

(b) If the given series for H is a central series, then $[H_i, H] \leq H_{i-1}$, so $[K_i, K] \leq K_{i-1}$. The proof of part (b) is now completed in a similar way to that of part (a). $\square$

Exercise Let G be a connected subgroup of $GL(n, F)$, so $G = G^0$. Prove that its derived subgroup G' is also connected but need not be closed in G.

For generalizations of all the above on the Zariski topology to more general things than subgroups of $GL(n, F)$, for further results on the Zariski topology on linear groups and also for a solution to the above exercise, see Chaps. 5 and 6 of Wehrfritz (1973a).

Chapter 5
Further Group-Theoretic Properties of Polycyclic Groups

In this chapter we continue our exposition from Chap. 2, but now we can make use of techniques developed in Chaps. 3 and 4.

We have already introduced the Frattini subgroup $\Phi(G)$ of a group G as the intersection of all the maximal subgroups of G, meaning G itself if none such exist. Also we proved in 1.17 that if G is finite then $\Phi(G)$ is nilpotent. Ito and Hirsch extended this as follows.

5.1 (Ito 1953; Hirsch 1954) If G is a polycyclic-by-finite group, then $\Phi(G)$ is nilpotent.

Proof Let N be a normal subgroup of $\Phi(G)$ of finite index. From 2.9 it follows that there is a subgroup K of G of finite index with $K \cap \Phi(G) = N$. Set $L = K_G = \bigcap_{g \in G} K^g$. Then L is normal of finite index in G and $L \cap \Phi(G) \leq N$. If H is any group and if θ is any homomorphism of H, then $\Phi(H)\theta \leq \Phi(H\theta)$. Hence $\Phi(G)L/L \leq \Phi(G/L)$, which we know is nilpotent. Thus $\Phi(G)/(L \cap \Phi(G))$ is nilpotent and, since $L \cap \Phi(G) \leq N$, $\Phi(G)/N$ too is nilpotent. Consequently $\Phi(G)$ is nilpotent by 2.8. $\qquad\square$

Exercise Let $G \in \mathbf{PF}$. Suppose $N \geq \Phi(G)$ is a normal subgroup of G with $N/\Phi(G)$ nilpotent. Prove that N is nilpotent. (You need to prove first the case where G is finite. Use Sylow's Theorem.)

Let G be polycyclic-by-finite. By a theorem of Merzljakov (1970), see comments after 4.10, there exists an integer n and an embedding of $\operatorname{Aut} G$ into $GL(n, \mathbf{Z})$. Hence by 4.4 any soluble subgroup of $\operatorname{Aut} G$ is polycyclic. This conclusion does not need the powerful result of Merzljakov.

5.2 (Smirnov 1953; Baer 1955a) Let Γ be a soluble group of automorphisms of the polycyclic-by-finite group G. Then Γ is polycyclic.

B.A.F. Wehrfritz, *Group and Ring Theoretic Properties of Polycyclic Groups*,
Algebra and Applications 10,
DOI 10.1007/978-1-84882-941-1_5, © Springer-Verlag London Limited 2009

Proof Now G has a characteristic series $\langle 1 \rangle = G_0 \le G_1 \le \cdots \le G_r = G$ such that each G_i/G_{i-1} is either finite or free abelian of finite rank (take the derived series of a characteristic polycyclic subgroup of G of finite index and then insert the maximal periodic subgroups in the abelian factors of the series). Thus each is normalized by Γ. Set $C_i = C_\Gamma(G_i/G_{i-1})$ and $C = \bigcap_i C_i$. If G_i/G_{i-1} is finite then Γ/C_i is finite soluble. If G_i/G_{i-1} is free abelian of rank n_i, then Γ/C_i embeds into $GL(n_i, \mathbf{Z})$ and therefore is polycyclic by 4.4. Therefore Γ/C is polycyclic. Now C stabilizes the series above, so C is finitely generated by 4.4.1 and nilpotent by 1.19. Thus C is polycyclic by 2.13 and consequently Γ is polycyclic. □

5.3 If N is a non-trivial normal subgroup of the nilpotent group G, then $N \cap \zeta_1(G) \ne \langle 1 \rangle$.

Proof Now $G = \zeta_c(G)$ for some c, so there exists a least i such that $N \cap \zeta_i(G) \ne \langle 1 \rangle$. Then $[G, N \cap \zeta_i(G)] \le N \cap \zeta_{i-1}(G) = \langle 1 \rangle$. Thus $\zeta_1(G) \ge N \cap \zeta_i(G) \ne \langle 1 \rangle$. □

5.4 Let G be a nilpotent group. If A is a maximal abelian normal subgroup of G, then $C_G(A) = A$.

Proof Now G/A is nilpotent. If $C_G(A) \ne A$, then by 5.3 there is some xA in $(C_G(A)/A) \cap \zeta_1(G/A) \backslash \langle 1 \rangle$. Then $\langle x, A \rangle$ is abelian, normal in G and larger than A. This contradiction of the maximality of A shows that $C_G(A) = A$. □

5.5 Let G be a nilpotent group. If every abelian normal subgroup of G is finitely generated, then G is polycyclic.

Proof By Zorn's Lemma (1.11) there is a maximal abelian normal subgroup A of G. By hypothesis A is polycyclic. Hence by 5.2 the group $G/C_G(A)$ (which embeds into Aut A note) is polycyclic. Consequently so too is G by 5.4. □

We cannot replace G nilpotent in 5.5 by G soluble. For example let G be the central product of infinitely many copies of the group H, amalgamating their centres, where H is the split extension of the quaternion group Q of order 8 by $\langle \phi \rangle$ and ϕ is an automorphism of Q of order 3. The centre of G (of order 2) is the only non-trivial abelian normal subgroup of G. This group G has derived length 3 and is not polycyclic.

Exercise If G is a metabelian group with each of its abelian normal subgroups finitely generated, prove that G is polycyclic.

5.6 (Baer R.) If G is a soluble group there exists a nilpotent normal subgroup B of G of class at most 2 with $C_G(B) = \zeta_1(B) \le B$.

Proof By Zorn's Lemma (1.11) there exists a maximal abelian normal subgroup A of G. If $A = C_G(A)$ set $B = A$. Suppose $A < C_G(A)$. Let B/A be an abelian normal subgroup of G/A that is maximal with respect to lying in $C_G(A)/A$; B exists by Zorn's Lemma. Clearly B is nilpotent of class at most 2.

Assume $C_G(B)$ is not contained in $B \geq A$. Then $C_G(B)/A$ contains a non-trivial abelian normal subgroup D/A of G/A. Then BD/A is abelian, contained in $C_G(A)/A$ and normal in G/A. By the choice of B we have $D \leq B$, so $D \leq B \cap C_G(B) = \zeta_1(B)$. But $A \leq \zeta_1(B)$, which is normal in G, so by the choice of A we have $A = \zeta_1(B)$ and $A = D$. This contradiction shows that $C_G(B) \leq B$. $\qquad\square$

5.7 Let G be a soluble group and assume that every abelian subnormal subgroup of G of (subnormal) depth at most 2 is finitely generated. Then G is polycyclic.

To say that X is subnormal in G of depth at most 2 just means that there exists Y with $X \triangleleft Y \triangleleft G$. The reader should have no difficulty in extending this definition to depth r.

Proof Pick B as in 5.6. By 5.5 the group B is polycyclic. Hence $G/C_G(B)$ is polycyclic by 5.2. Consequently G is polycyclic. $\qquad\square$

5.8 **Corollary** Let G be a soluble-by-finite group each of whose abelian subnormal subgroups is finitely generated. Then G is polycyclic-by-finite.

Exercise Prove 5.8 directly from 5.2 without using 5.5, 5.6 or 5.7.

5.9 **Corollary** (Mal'cev 1951) Let G be a soluble-by-finite group. Then G is polycyclic-by-finite if and only if every abelian subgroup of G is finitely generated.

Continuing the theme above, in a long and difficult paper Wilson (1982a) makes a deep study of the influence of the abelian subnormal subgroups A of an infinite polycyclic group G on the group G itself. For example he shows that if $h(A) \leq n$ for all such A, then $h(G) \leq (n^2 + 3n - 2)/2$ and $h(G/\eta_1(G)) \leq n - 1$. Further Wilson (1982b) considers the nilpotent subgroups H of our polycyclic group G and shows that if G is also torsion-free, then $h(H) \leq h(\eta_1(G))$ for all such H with equality implying $H \leq \eta_1(G)$.

Merzljakov (1969) has constructed a locally soluble group with all its abelian subgroups finitely generated that is not polycyclic. For those who know the terminology I mention the following.

If G is a radical (meaning hyper locally nilpotent) group with all its abelian subgroups finitely generated, then G is polycyclic (see Robinson 1972, 3.31).

If G is a hyperabelian group with all its abelian subnormal subgroups of depth at most 2 finitely generated, then G is polycyclic. (Modify the proofs of 5.6 and 5.7, cf. Kegel and Wehrfritz (1973), 1.A.8 and 1.G.2 and use Zassenhaus' Theorem that locally soluble linear groups are soluble, see Wehrfritz 1973a, 3.8).

Exercise Let G be a soluble-by-finite group with each of its normal subgroups finitely generated. Prove that G is polycyclic-by-finite.

Exercise Let G be a soluble-by-finite group with the maximal condition on subnormal subgroups of depth at most 2. Again prove that G is polycyclic-by-finite.

Note that $\mathbf{S} \cap \mathbf{Max\text{-}n}$ does not lie in $\mathbf{P}$. For by 3.10 the group $W = C\,wr\,C$, for C infinite cyclic, is finitely generated with max-n, while its base group is clearly not finitely generated. For further remarks in this area see after 3.11. As a companion piece to 5.5, note that $\mathbf{N} \cap \mathbf{Max\text{-}n} \subseteq \mathbf{P}$ by 3.11 and 2.13.

5.10 $\mathbf{G} \cap \mathbf{N} \cap \mathbf{AF} \subseteq \mathbf{FA}$.

Proof Consider $G \in \mathbf{G} \cap \mathbf{N} \cap \mathbf{AF}$ of class c. By 2.14(c) we may assume that G is torsion-free. Set $Z_i = \zeta_i(G)$. By 2.15 each Z_{i+1}/Z_i is torsion-free. Suppose $G = Z_2 > Z_1$. Let $x, y \in G$. There exists $r > 0$ such that $[x^r, y^r] = 1$. Since G is nilpotent of class 2 this yields $[x, y]^{2r} = 1$. But G is torsion-free. Thus $[x, y] = 1$ and G is abelian. Apply this to G/Z_{c-2}. We find that $G = Z_{c-1}$, an impossibility, so G is abelian. $\qquad\square$

Exercise Prove that $L\mathbf{N} \cap \mathbf{A}(L\mathbf{F}) \subseteq (L\mathbf{F})\mathbf{A}$.

If G is any group and if H is a subgroup of G, set

$$i_G(H) = \{g \in G\colon \text{some positive power of } g \text{ lies in } H\}.$$

It is called the *isolator* of H in G. It is not usually a subgroup of G. For example, $i_G(\langle 1 \rangle)$ is always the set of elements of G of finite order and if G is the infinite dihedral group D_∞ this set is not a subgroup. If $H = i_G(H)$ we say H is *isolated* (in G).

5.11   Let H be a subgroup of the finitely generated nilpotent group G. Then $i_G(H)$ is a subgroup of G and the index $(i_G(H) : H)$ is finite.

Proof By 2.14 there exist subgroups $H = H_0 \triangleleft H_1 \triangleleft \cdots \triangleleft H_r = G$ with each H_{i+1}/H_i finite or infinite cyclic. Suppose that H_{i+1}/H_i is finite and H_i/H_{i-1} is infinite cyclic. Then $H_i' \leq H_{i-1}$, $H_i' \triangleleft H_{i+1}$ and $H_{i+1}/H_i' \in \mathbf{FA}$ by 5.10. Let T/H_i' be the torsion subgroup of H_{i+1}/H_i'. Then H_{i-1} has finite index in $K = TH_{i-1}$ and the latter is normal in H_{i+1}, since H_{i+1}/T is abelian, with H_{i+1}/K infinite cyclic by finite. Thus there is a normal subgroup L of H_{i+1} with H_{i-1} subnormal in L, the index $(L : H_{i-1})$ finite and H_{i+1}/L infinite cyclic. Hence we may choose the H_i such that for some s the index $(H_s : H)$ is finite and for $i > s$ each H_i/H_{i-1} is infinite cyclic. Consequently $i_G(H) = H_s$, a subgroup of G of finite index over H. $\quad\square$

Exercise If H is a subgroup of the locally nilpotent group G, prove that $i_G(H)$ is a subgroup of G.

If π is any set of primes, set $i_G^\pi(H) = \{g \in G : g^r \in H$ for some positive π-number $r\}$.

5.12 Corollary Let H be a subgroup of the finitely generated nilpotent group G. Then $i_G^\pi(H)$ is a subgroup of G of finite index over H a π-number.

Proof Let $K = i_G(H)$. By 5.11 the index $(K : H)$ is finite, so $(K : H_K)$ is also finite (here $H_K = \bigcap_{k \in K} H^k$). Then K/H_K is a finite nilpotent group. Let T/H_K denote the maximal π-subgroup of K/H_K. Then $i_G^\pi(H) = HT$. The claims follow. $\square$

Exercise If H is a subgroup of the locally nilpotent group G, prove that $i_G^\pi(H)$ is a subgroup of G.

The example of the infinite dihedral group D_∞ shows that these isolator results break down even for abelian-by-finite polycyclic groups. However we do have the following (see Rhemtulla and Wehrfritz 1984).

Let G be a polycyclic group. Then G has a normal subgroup N of finite index such that for every subgroup H of N the isolator $i_N(H)$ is a subgroup of N of finite index over H. Moreover if G is a finitely generated soluble group and if G has a subgroup N of finite index such that if $H \leq N$ then $i_N(H)$ is a subgroup, that is of finite index over H whenever H/H_N is finitely generated, then G must be polycyclic. The proofs depend crucially on the Auslander Embedding Theorem (see 4.8). We do need the final hypothesis, for the finitely generated soluble group

$$G = \left\langle \begin{pmatrix} 1 & 0 \\ 1 & 1 \end{pmatrix}, \begin{pmatrix} 2 & 0 \\ 0 & 1 \end{pmatrix} \right\rangle$$

has the property that $i_G(H)$ is a subgroup of G for every subgroup H of G, but G is not polycyclic-by-finite.

5.13 Let X be a subset of the locally nilpotent group G. If G is torsion-free, then $C_G(X)$ is isolated in G. If H is a subgroup of G then $i_G(N_G(H)) \subseteq N_G(i_G(H))$. If also G is finitely generated (and so nilpotent), then $i_G(N_G(H)) = N_G(i_G(H))$.

For proofs see Hall (1958), pp. 25 and 26 or Lennox and Robinson (2004), 2.3.7 and 2.3.8. We now prove the following special case of 2.22, the conjugacy separability of polycyclic groups, which we left without proof.

5.14 Theorem (Blackburn 1965) If G is a finitely generated nilpotent group then G is conjugacy separable.

Proof Suppose a and b are elements of G that are not conjugate in G. We have to find a normal subgroup N of G of finite index such that aN and bN are not conjugate in G/N. Suppose a^G is (profinitely) closed in G. Now $b \notin a^G$, so there exists N normal of finite index in G with $bN \cap a^G = \emptyset$. In this case aN and bN are not conjugate in G/N. Thus we need to prove that a^G is closed in G.

If $\zeta_1(G)$ is finite, then so is G by 2.15 and the result is clear. Hence assume that G contains an infinite cyclic, central subgroup Z. Set $C = C_G(aZ/Z)$. Now $c \mapsto [a, c]$ is a homomorphism of C into Z. Thus $a^C = a[a, C]$, and $[a, C]$ being a subgroup of C is closed in C by 2.24. Hence a^C is closed in C. By induction on the Hirsch number of G we may assume for each $k = 1, 2, \ldots$ that $a^G Z^k$ is closed in G.

Let $b \notin a^G$ lie in the closure of a^G in G. Since $a^G Z$ is closed we have $b \in a^G Z$, say $b^g = az$ with $g \in G$ and $z \in Z$. Then $b^g \in C$. If $b^g \in a^C$, then $b \in a^G$, which is false. Thus $b^g \notin a^C$, which is closed in C. Hence $b^g M \cap a^C = \emptyset$ for some M normal of finite index in C. Set $k = (C : M)$. Then $Z^k \leq M$ and $b^g Z^k \cap a^C = \emptyset$. If $b^g Z^k \cap a^G \neq \emptyset$, we have $a^x = b^g y$ for some $x \in G$ and $y \in Z^k$. Also $b^g = az$, so $a^{-1}a^x = zy \in Z$ and so $x \in C$ by definition of C. Thus $b^g Z^k \cap a^C \neq \emptyset$, a contradiction, so $b^g Z^k \cap a^G = \emptyset$. Hence $b \notin a^G Z^k$, which contradicts the fact that $a^G Z^k$ is closed in G. Consequently no such b exists and the theorem is proved. $\square$

The proof of 2.22 has the same general layout as the proof of 5.14 above, but the details are considerably more involved. Also it requires the following non-trivial piece of number theory due to Schmidt (1930) and Chevalley (1951): if R is the ring of algebraic integers in a finite extension field of the rationals $\mathbf{Q}$ and if V is a subgroup of the group of units U of R of finite index, then there is a positive integer n such that $U \cap (1 + nR) \leq V$.

The matrix ring $\mathbf{Z}^{n \times n}$ is additively a free abelian group of finite rank n^2. As such it carries its own profinite topology, called the congruence topology. Left and right multiplication of elements on $\mathbf{Z}^{n \times n}$ are continuous. Also the topology is generated by ideals, for $\{r\mathbf{Z}^{n \times n} : r = 1, 2, \ldots\}$ is a base of the open neighborhoods of 0 in $\mathbf{Z}^{n \times n}$. It is a celebrated theorem of Mennicke (1965) that for $n \geq 3$ the congruence topology induces on $GL(n, \mathbf{Z})$ its profinite topology. (This is false for $n = 2$ and trivial for $n = 1$.) The following related result depends on conjugacy separability.

5.15 (Wehrfritz 1973c) Let G be a soluble-by-finite subgroup of $GL(n, \mathbf{Z})$. Then G is closed in the congruence topology and the congruence topology induces on G the profinite topology of G.

Proof Now $\mathbf{Z}^{n \times n}$ is a G-module via right multiplication. Set $H = G[\mathbf{Z}^{n \times n}$, the split extension of $\mathbf{Z}^{n \times n}$ by G. Then H is polycyclic-by-finite by 4.4. Hence H is conjugacy separable by 2.22. Now the conjugacy class of $1_n \in \mathbf{Z}^{n \times n}$ in H is $1_n G = G$ as a subset of $\mathbf{Z}^{n \times n}$. Also the profinite topology of H induces on $\mathbf{Z}^{n \times n}$ the latter's profinite topology (see 2.24), that is, induces the congruence topology. Therefore G is closed in the latter.

Let N be a normal subgroup of G of finite index. Then by the above N is closed in the congruence topology on $\mathbf{Z}^{n \times n}$ and therefore in G. Hence xN is congruence closed in G for all $x \in G$ and consequently $G \setminus N$ is closed and N is open in this topology. Therefore profinite open subsets of G are also congruence open. The converse is obvious. $\square$

5.16 Theorem (Roseblade 1978) Let G be a polycyclic-by-finite group. Then there exists a normal subgroup M of G of finite index such that if H is a subgroup of M with $(M : N_M(H))$ finite, then H^M/H_M is finite.

Recall that $H^M = \langle H^x : x \in M \rangle$ and $H_M = \bigcap_{x \in M} H^x$, both of which are normal in M. Roseblade names this property of M *orbitally sound*. It is an important ingredient in his study of the prime ideals in the group ring of a polycyclic group, see Chap. 8 below. This is another example where we can use the Auslander Embedding Theorem to give a short proof and in fact it is a special case of 8.6 below. Since we have no need to use 5.16 until Chap. 8 we postpone its proof until then. Roseblade's original proof in Roseblade (1978) uses the techniques we have been using above.

This chapter will complete our discussion of the purely group theoretic properties of polycyclic groups, the remaining chapters being devoted to their ring and module theoretic properties. We conclude this chapter by just summarizing a few more group theoretic properties of polycyclic groups.

We begin with a few further theorems involving the profinite topology. Jeanes and Wilson (1978) prove the following. Let G be a finitely generated soluble-by-finite group (actually a weaker hypothesis will suffice). If every subgroup of G that is subnormal in G of depth at most 2 is (profinitely) closed in G, then G is polycyclic-by-finite. This is a sort of converse to Mal'cev's Theorem 2.9. The proof is quite straightforward and uses the techniques of Chap. 3.

Let G be a soluble-by-finite group and M a finitely $\mathbf{Z}$-generated G-module. Grunewald and Segal (1978), see also Segal (1983), p. 65, prove that if $\delta : G \to M$ is a *derivation* of G to M (meaning that $(xy)\delta = x\delta.y + y\delta$ for all $x, y \in G$), then $G\delta$ is profinitely closed in M. This can be proved as a corollary of 2.22 (conjugacy separability of **PF**-groups). In turn it can also be used to prove the following analogue of 2.22. Grunewald and Segal (1978): let G be a polycyclic-by-finite group and H and K subgroups of G. Then H and K are conjugate in G if and only if HN/N and KN/N are conjugate in G/N for every normal subgroup N of finite index in G.

Nikolov and Segal (2007) use profinite groups to prove that if N is a normal subgroup of a polycyclic-by-finite group G with G isomorphic to the direct product of N and G/N, then N is a direct factor of G. Again using profinite groups and also Lie algebras, Mann and Segal (2007) study the breadth $b(G)$ of a polycyclic group G, where

$$b(G) = \max\{h(G) - h(C_G(x)) : x \in G\}.$$

In particular they prove that $b(G) + 1$ bounds the class of G if G is also nilpotent and in general that $h(G')$ is bounded by a function of $b(G)$.

We have already recorded, see 2.21, Lennox and Wilson's result on the product of two subgroups in a polycyclic-by-finite group and remarked (see p. 27) on its relation to Lennox and Wilson (1979). We have also mentioned Lennox & Roseblade's result 3.12. More in this vein is known. Almazar and Cossey (1996) prove amongst other things that if $G = AB$ is polycyclic with A and B nilpotent subgroups of G, then

$$h(G) \le h(A) + h(B).$$

Rhemtulla and Wilson (1988) characterize the subgroup H, subnormal in some subgroup of finite index in a polycyclic-by-finite group G, in terms of products of H with other subgroups of G.

Linnell and Warhurst (1981) consider the minimal number $d(G)$ of generators of the polycyclic-by-finite group G. They prove that $d(G) \leq 1 + d(G/N)$ for at least one normal subgroup N of G of finite index. The proof uses lattices over orders.

Robinson (2002) studies the permutability of subgroups in a polycyclic-by-finite group, using derivations (see above for the definition of derivation). Using concepts coming from topology or more particularly from K-theory, Moravec (2007) shows that the (non-abelian) tensor product of two polycyclic groups is again polycyclic. Also Cossey (1991) studies the Wielandt subgroup $w(G)$ of a polycyclic group G; here $w(G)$ denotes the intersection of the normalizers of the subnormal subgroups of G. For example he shows that $C_G(w(G))$ has finite index in G.

For an excellent introduction to arithmetic methods for studying polycyclic groups, see Segal's book (1983). For some more recent work in this area, see du Sautoy (2002) and Baues and Grunewald (2006). For algorithmic and decidable properties of polycyclic groups see Segal (1990) and Baumslag et al. (1991). See also Sims (1994) Chap. 9 and Holt et al. (2005) Chap. 8.

Chapter 6
Hypercentral Groups and Rings

The theme of this chapter is the interplay between group-theoretic nilpotence properties of a group G and ring-theoretic nilpotence properties of its group ring $\mathbf{Z}G$. Our main aim is to prove Roseblade's theorems that $\mathbf{Z}G$ is a hypercentral ring if and only if G is a hypercentral group and that $\mathbf{Z}G$ is a polycentral ring if and only if G is a finitely generated nilpotent group. We must start by explaining these terms.

An ascending chain $\{G_\alpha\}_{\alpha<\gamma}$ of subgroups of G is an *ascending series* if each G_α is normal in $G_{\alpha+1}$ and if $G_\lambda = \bigcup_{\alpha<\lambda} G_\alpha$ for all limit ordinals $\lambda < \gamma$. For example, the upper central series $\{\zeta_\alpha(G)\}$, continued transfinitely in the obvious way, is an ascending series of G. Its union is the *hypercentre* of G, which we denote by $\zeta(G)$; that is $\zeta(G) = \bigcup_\alpha \zeta_\alpha(G)$. A group G is *hypercentral* if and only if $G = \zeta(G)$. The following result we leave as an exercise for the reader. It gives equivalent formulations of the definition of a hypercentral group.

6.1 The following are equivalent for the group G.

(a) $G = \zeta_\gamma(G)$ for some ordinal γ.
(b) There is a series $\langle 1 \rangle = G_0 \le G_1 \le \cdots \le G_\alpha \le \cdots \le G_\gamma = G$ with $[G_{\alpha+1}, G] \le G_\alpha$ for all $\alpha < \gamma$.
(c) If $N \ne G$ is a normal subgroup of G, then $\zeta_1(G/N) \ne \langle 1 \rangle$.
(d) If $M < N$ are normal subgroups of G, then $N/M \cap \zeta_1(G/M) \ne \langle 1 \rangle$.

6.2 Let G be a finitely generated hypercentral group. Then G is nilpotent.

Proof Let $G = \langle g_1, g_2, \ldots, g_n \rangle$. If $x \in \zeta_{\omega+1}(G)$, then $[x, g_j] \in \zeta_\omega(G) = \bigcup_{i<\omega} \zeta_i(G)$. Since n is finite there exists $r < \omega$ with $[x, g_j] \in \zeta_r(G)$ for each j. Thus x is central modulo $\zeta_r(G)$ and consequently $x \in \zeta_{r+1}(G)$. But then $\zeta_{\omega+1}(G) = \zeta_\omega(G)$, so $G = \zeta_\omega(G)$. Therefore there is some $s < \omega$ with each g_j in $\zeta_s(G)$. Consequently $G = \zeta_s(G)$, which is nilpotent. $\qquad\square$

The correct (that is to say, useful) notion of a hypercentral ring is not quite so obvious. The centre of a ring is seldom an ideal and apart from trivial cases a ring

B.A.F. Wehrfritz, *Group and Ring Theoretic Properties of Polycyclic Groups*, Algebra and Applications 10, DOI 10.1007/978-1-84882-941-1_6, © Springer-Verlag London Limited 2009

has very few of its ideals in its centre. The correct analogue of a central subgroup turns out to be an ideal generated as an ideal by central elements.

An ideal $\mathbf{a}$ of a ring R is a *hypercentral ideal* if it has an ascending series

$$\{0\} = \mathbf{a}_0 \leq \mathbf{a}_1 \leq \cdots \leq \mathbf{a}_\alpha \leq \cdots \leq \mathbf{a}_\gamma = \mathbf{a}$$

of ideals of R such that for each $\alpha < \gamma$ the ideal $\mathbf{a}_{\alpha+1}/\mathbf{a}_\alpha$ is generated as an ideal by central elements of the ring $R/\mathbf{a}_\alpha$. Further $\mathbf{a}$ is a *polycentral ideal* of R if there is a finite series

$$\{0\} = \mathbf{a}_0 \leq \mathbf{a}_1 \leq \cdots \leq \mathbf{a}_i \leq \cdots \leq \mathbf{a}_r = \mathbf{a}$$

of ideals of R such that each ideal $\mathbf{a}_{i+1}/\mathbf{a}_i$ is generated by finitely many central elements of $R/\mathbf{a}_i$. The ring R is *hypercentral* (resp. *polycentral*) if each of its ideals (including R itself) is hypercentral (resp. polycentral).

If ϕ is a ring homomorphism of R, clearly $\mathbf{a}\phi$ is hypercentral, resp. polycentral, in $R\phi$ whenever $\mathbf{a}$ is hypercentral, resp. polycentral, in R. In particular we have the following.

6.3   The ring R is hypercentral if and only if whenever $\mathbf{a} < \mathbf{b}$ are ideals of R the factor $\mathbf{b}/\mathbf{a}$ contains a non-zero central element of $R/\mathbf{a}$. Also R is polycentral if and only if R is hypercentral with all of its ideals finitely generated.

6.4   **Lemma** (Roseblade and Smith 1976) Let J be a ring, X a subset of J, G a group, H a group acting on G and N a normal subgroup of G for which there is an ascending series

$$N = G_0 \lhd G_1 \lhd \cdots \lhd G_\lambda \lhd G_{\lambda+1} \lhd \cdots \lhd G_\rho = G$$

of H-invariant subgroups of G. Suppose that for any H-ideals (i.e. H-invariant ideals) $\mathbf{a} < \mathbf{b}$ of JN there exists $\alpha \in \mathbf{b}\backslash\mathbf{a}$ such that $\alpha^h - \alpha$ and $\alpha x - x\alpha$ lie in $\mathbf{a}$ for all $h \in H$ and $x \in X$ (that is, suppose JN is 'hyper X-H central'). Then for any H-ideals $\mathbf{a} < \mathbf{b}$ of JG there exists a finite set $\lambda(1), \lambda(2), \ldots, \lambda(m)$ of ordinals and an α in $\mathbf{b}\backslash\mathbf{a}$ such that $\alpha^k - \alpha$ and $\alpha x - x\alpha$ lie in $\mathbf{a}$ for all k in $\bigcap_{1 \leq i \leq m} C_H(G_{\lambda(i)+1}/G_{\lambda(i)})$ and x in X.

(If $X = \emptyset$ this is Lemma 7 of Roseblade and Smith (1976).)

Proof   We induct on ρ; if $\rho = 0$ there is nothing to prove. If ρ is a limit ordinal, then $G = \bigcup_{\tau<\rho} G_\tau$ and there exists $\tau < \rho$ with $\mathbf{a} \cap JG_\tau < \mathbf{b} \cap JG_\tau$. The result then follows by induction. Hence assume $\rho = \tau + 1$.

Set $S = JG_\tau$. Since $JG = \sum_{g \in G} Sg$ there is a finite set of elements $g_1, g_2, \ldots,$ g_n of G such that

$$\mathbf{a} \cap (Sg_1 + Sg_2 + \cdots + Sg_n) < \mathbf{b} \cap (Sg_1 + Sg_2 + \cdots + Sg_n).$$

Choose $g_1, \ldots, g_n$ so that n is minimal. By multiplying through by g_1^{-1} we may assume that $g_1 = 1$. Set $T = Sg_2 + Sg_3 + \cdots + Sg_n$, $\mathbf{a}_1 = S \cap (\mathbf{a} + T)$ and $\mathbf{b}_1 =$

$S \cap (\mathbf{b} + T)$. Since G_τ is normal in G, so $\mathbf{a}_1$ and $\mathbf{b}_1$ are ideals of S. By construction $\mathbf{a}_1 < \mathbf{b}_1$, for suppose $\mathbf{a}_1 = \mathbf{b}_1$. By the choice of n we have $\mathbf{a} \cap T = \mathbf{b} \cap T$; also

$$(\mathbf{a} \cap (S + T)) + T = (\mathbf{a} + T) \cap (S + T) = (S \cap (\mathbf{a} + T)) + T = \mathbf{a}_1 + T = \mathbf{b}_1 + T$$

$$= \mathbf{b} \cap (S + T)) + T$$

in the same way. Consequently

$$\mathbf{b} \cap (S + T) = (\mathbf{a} \cap (S + T)) + (\mathbf{b} \cap (S + T) \cap T) = \mathbf{a} \cap (S + T),$$

which is false.

Clearly $\mathbf{a}_1$ and $\mathbf{b}_1$ are $C_H(G/G_\tau)$-invariant, so by induction there are finitely many ordinals, say $\lambda(2), \ldots, \lambda(m)$ and an α_1 in $\mathbf{b}_1 \backslash \mathbf{a}_1$ such that $\alpha_1^k - \alpha_1$ and $\alpha_1 x - x\alpha_1$ lie in $\mathbf{a}_1$ for all k in $K = C_H(G/G_\tau) \cap \bigcap_{2 \le i \le m} C_H(G_{\lambda(i)+1}/G_{\lambda(i)})$ and for all x in X. There exists $t \in T$ with $\alpha = \alpha_1 + t \in \mathbf{b}$ and if $\alpha \in \mathbf{a}$, then $\alpha_1 \in S \cap (\mathbf{a} + T) = \mathbf{a}_1$, so $\alpha \notin \mathbf{a}$.

For any $k \in K$ we have

$$\alpha^k - \alpha \in \mathbf{b} \cap (\alpha_1^k - \alpha_1 + T), \quad \text{since } [G, K] \le G_\tau \subseteq S,$$

$$\le \mathbf{b} \cap (\mathbf{a}_1 + T) \quad \text{by the choice of } \alpha_1,$$

$$= \mathbf{a}_1 + (\mathbf{b} \cap T) = \mathbf{a}_1 + (\mathbf{a} \cap T) \le \mathbf{a}.$$

Finally for all $x \in X$ we have $\alpha x - x\alpha \in \mathbf{b} \cap (\mathbf{a}_1 + T) \le \mathbf{a}$, and the proof is complete. $\qquad\square$

Exercise (See 5.3.5 of Shirvani and Wehrfritz 1986) Let J be a ring, G a group, H a group acting on G and N a subgroup of G for which there is an ascending series

$$N = G_0 \lhd G_1 \lhd \cdots \lhd G_\lambda \lhd G_{\lambda+1} \lhd \cdots \lhd G_\rho = G$$

of H-invariant subgroups of G. Suppose that

(a) $(H : C_H(G_{\lambda+1}/G_\lambda))$ is finite for all $\lambda < \rho$ and
(b) for any H-ideals $\mathbf{a} < \mathbf{b}$ of JN there exists α in $\mathbf{b} \backslash \mathbf{a}$ with the index in H of $C_H(\alpha + \mathbf{a}/\mathbf{a})$ finite.

Prove that for any H-ideals $\mathbf{a} < \mathbf{b}$ of JG there exists α in $\mathbf{b} \backslash \mathbf{a}$ with the index in H of $C_H(\alpha + \mathbf{a}/\mathbf{a})$ is finite.

6.5   Corollary (Roseblade 1971) If J is a hypercentral ring and G is a hypercentral group, then JG is a hypercentral ring.

In particular if G is a hypercentral group, then $\mathbf{Z}G$ is a hypercentral ring. The converse will take a fair bit of further work.

Proof Let $\mathbf{a} < \mathbf{b}$ be ideals of JG. Apply 6.4 with $X = J$, $H = G$ acting on itself by conjugation, $N = \langle 1 \rangle$ and the series $\{G_\lambda\}$ the upper central series of G. By hypothesis $J = JN$ is a hypercentral ring (and hence is hyper X-H central here). Also $G = \bigcap_\lambda C_G(G_{\lambda+1}/G_\lambda)$. Thus there exists an α in $\mathbf{b} \backslash \mathbf{a}$ such that $g^{-1}\alpha g - \alpha$ and $\alpha x - x\alpha$ lie in $\mathbf{a}$ for all g in G and x in J. That is, α is central modulo $\mathbf{a}$ in JG and the proof is complete. $\qquad\square$

6.6   Corollary (Roseblade and Smith 1976) Let J be a ring, G a group and M a normal subgroup of G that is hypercentral as a group. If $\mathbf{a} < \mathbf{b}$ are ideals of JG, there exists α in $\mathbf{b} \backslash \mathbf{a}$ such that $\alpha^k - \alpha \in \mathbf{a}$ for every $k \in M$.

Proof Let $\langle 1 \rangle = G_0 \leq G_1 \leq \cdots \leq G_\lambda \leq \cdots \leq G_\tau = M$ be the upper central series of M and set $G_{\tau+1} = G$. Let $H = G$, again acting by conjugation, $N = \langle 1 \rangle$ and $X = \emptyset$. Each G_λ is normal in G. By 6.4 there exists some α in $\mathbf{b} \backslash \mathbf{a}$ such that $\alpha^k - \alpha \in \mathbf{a}$ for every k in $\bigcap_{\lambda \leq \tau} C_G(G_{\lambda+1}/G_\lambda) = K$ say. Clearly $M \leq K$. $\qquad\square$

Intersection theorems play quite a big role in the work of Roseblade, Zalesskii and others on group rings. Here is an example.

6.7   Corollary Let J, G and M be as in 6.6 and set

$$\Delta_G(M) = \{g \in G: \text{ the index } (M : C_M(g)) \text{ is finite}\}.$$

Then $\mathbf{b} \cap J\Delta_G(M) \neq \{0\}$ for every non-zero ideal $\mathbf{b}$ of JG.

Exercise Prove that $\Delta_G(M)$ is a normal subgroup of G.

Proof By 6.6 (with $\mathbf{a} = \{0\}$) there exists $\alpha \in \mathbf{b} \backslash \{0\}$ such that $\alpha^g = \alpha$ for every $g \in M$. Suppose $\alpha = \alpha_1 g_1 + \alpha_2 g_2 + \cdots + \alpha_r g_r$ for non-zero elements α_i of J and distinct elements g_i of G. Then

$$\alpha_1 g_1^g + \alpha_2 g_2^g + \cdots + \alpha_r g_r^g = \alpha_1 g_1 + \alpha_2 g_2 + \cdots + \alpha_r g_r$$

for every g in M. Hence

$$g_1^M \cup g_2^M \cup \cdots \cup g_r^M = \{g_1, g_2, \ldots, g_r\}.$$

Thus $(M : C_M(g_i)$ is finite for each i and therefore the g_i lie in $\Delta_G(M)$. Consequently $\alpha \in \mathbf{b} \cap J\Delta_G(M)$ and so $\mathbf{b} \cap J\Delta_G(M) \neq \{0\}$. $\qquad\square$

The Artin-Rees Properties

The term Artin-Rees is used for a number of very closely related properties and one has to check carefully when reading which particular version of the Artin-Rees property the author is assuming. In this book we shall use the following terminology. Let R be a ring (with 1 as always) and let $\mathbf{a}$ be an ideal of R.

$\mathbf{a}$ is *right strong Artin-Rees* if the subring $\langle R, \mathbf{a}X \rangle = \bigoplus_{0 \le i < \infty} \mathbf{a}^i X^i$ of the polynomial ring $R[X]$ in the one variable X is right Noetherian.

$\mathbf{a}$ is *right Artin-Rees* if for each finitely generated right R-module M and each submodule N of M there exists an integer $k \ge 0$ such that for all $n \ge k$

$$N \cap M\mathbf{a}^n = (N \cap M\mathbf{a}^k)\mathbf{a}^{n-k}, \quad \le N\mathbf{a}^{n-k} \text{ note.}$$

$\mathbf{a}$ is *right weak Artin-Rees* if for each M and N as above there exists an integer $m \ge 0$ with $N \cap M\mathbf{a}^m \le N\mathbf{a}$.

The ring R is right strong Artin-Rees if each of its ideals is right strong Artin-Rees, and similarly with the other Artin-Rees variants.

There is a fourth Artin-Rees property between right Artin-Rees and right weak Artin-Rees, namely that the $\mathbf{a}$-adic topology on M induces on N the $\mathbf{a}$-adic topology on N. The reader will probably be relieved to be told that we have no need of this one. We will, however, make use of the other three. Authors requiring only one version of Artin-Rees tend to call it the Artin-Rees property, regardless of how strong or weak it is, so readers need to be cautious. There are obvious left analogues of these right Artin-Rees properties.

6.8 Right strong Artin-Rees implies right Artin-Rees, which implies right weak Artin-Rees.

Proof Suppose $R^\wedge = \langle R, \mathbf{a}X \rangle = R \oplus \mathbf{a}X \oplus \mathbf{a}^2 X^2 \oplus \cdots$ is right Noetherian. Set

$$M^\wedge = \bigoplus_{i \ge 0} M\mathbf{a}^i X^i = M \oplus M\mathbf{a}X \oplus M\mathbf{a}^2 X^2 \oplus \cdots \oplus M\mathbf{a}^i X^i \oplus \cdots .$$

Then $M^\wedge$ is a finitely generated $R^\wedge$-module (it is easy to check that $M^\wedge = MR^\wedge$) and so is $R^\wedge$-Noetherian. For N a R-submodule of our right R-module M, set

$$N^\sim = \bigoplus_{i \ge 0} (N \cap M\mathbf{a}^i) X^i = N \oplus (N \cap M\mathbf{a}) X \oplus (N \cap M\mathbf{a}^2) X^2 \oplus \cdots .$$

Then $N^\sim$ is an $R^\wedge$-submodule of $M^\wedge$, so $N^\sim$ is finitely generated. Thus for some $k \ge 0$

$$N^\sim = (N \oplus (N \cap M\mathbf{a}) X \oplus \cdots \oplus (N \cap M\mathbf{a}^k) X^k) R^\wedge.$$

Equate coefficients of X^n. Thus for all $n \ge k$ we have

$$N \cap M\mathbf{a}^n = N\mathbf{a}^n + (N \cap M\mathbf{a})\mathbf{a}^{n-1} + \cdots + (N \cap M\mathbf{a}^k)\mathbf{a}^{n-k}.$$

Now for $i < k \le n$ we have

$$(N \cap M\mathbf{a}^i)\mathbf{a}^{n-i} = (N \cap M\mathbf{a}^i)\mathbf{a}^{k-i}\mathbf{a}^{n-k} \le (N \cap M\mathbf{a}^k)\mathbf{a}^{n-k}.$$

Therefore $N \cap M\mathbf{a}^n = (N \cap M\mathbf{a}^k)\mathbf{a}^{n-k}$, as required.

Finally suppose $\mathbf{a}$ is right Artin-Rees. Thus given M and N we have k so that $N \cap M\mathbf{a}^n = (N \cap M\mathbf{a}^k)\mathbf{a}^{n-k}$ for $n \geq k$. Set $m = k + 1$. Then $N \cap M\mathbf{a}^m = (N \cap M\mathbf{a}^k)\mathbf{a} \leq N\mathbf{a}$. The proof is complete. $\qquad\square$

6.9   Theorem (Roseblade 1976) Let $\langle S, G \rangle$ be a ring generated as a ring by its right Noetherian subring S and the polycyclic-by-finite subgroup G of its group of units with G normalizing S (e.g. $R = SG$). Let $\mathbf{a}$ be a centrally generated ideal of S normalized by G. Then $\mathbf{a}G = G\mathbf{a}$ is a right strong Artin-Rees ideal of R.

Given a ring $R = \langle S, G \rangle$ as above with G normalizing S we will frequently copy polynomial ring notation and write $R = S[G]$.

Note that in 6.9, since G normalizes S and $\mathbf{a}$, the right ideal $\mathbf{a}G$ is an ideal of R. Thus $\mathbf{a}G = \mathbf{a}R = G\mathbf{a} = R\mathbf{a}$. If M is a right R-module, then $M(\mathbf{a}G)^r = MG\mathbf{a}^r = M\mathbf{a}^r$, so we can replace $\mathbf{a}G$ by $\mathbf{a}$ in the formula for the right Artin-Rees property.

Proof In the polynomial ring $R[X]$ we have the subring

$$\langle R, \mathbf{a}GX \rangle = \langle S, G, \mathbf{a}X \rangle = \langle \langle S, \mathbf{a}X \rangle, G \rangle.$$

Now $\langle S, \mathbf{a}X \rangle = \langle S, z_1 X, z_2 X, \ldots, z_r X \rangle$, where the z_i form a finite central generating set for $\mathbf{a}$; we are using here that S is right Noetherian. By the Hilbert Basis Theorem (or use 3.7.1) the polynomial ring $S[X_1, X_2, \ldots, X_r]$ is right Noetherian. Thus $\langle S, \mathbf{a}X \rangle$ is also right Noetherian. Consequently so is $\langle S, \mathbf{a}X, G \rangle$, see 3.7.3. Therefore $\mathbf{a}$ is right strong Artin-Rees. $\qquad\square$

6.10   Corollary Let R be a right Noetherian ring. Then any centrally generated ideal of R is right strong Artin-Rees.

Proof Just take $G = \langle 1 \rangle$ in 6.9. $\qquad\square$

6.11   Corollary Let G be a polycyclic-by-finite group, let A be an abelian normal subgroup of G and let $\mathbf{a}$ be any ideal of $\mathbf{Z}A$ normalized by G. Then $\mathbf{a}G$ is a right strong Artin-Rees ideal of $\mathbf{Z}G$.

Proof Take $R = \mathbf{Z}G$ and $S = \mathbf{Z}A$ in 6.9. $\qquad\square$

We need to cope not just with centrally generated ideals but with polycentral ideals. Here our conclusions are slightly weaker.

6.12   Theorem (Roseblade 1976) Let $\langle S, G \rangle$ be a ring generated as a ring by its right Noetherian subring S and the polycyclic-by-finite subgroup G of its group of units with G normalizing S. If $\mathbf{a}$ is a polycentral ideal of S normalized by G, then $\mathbf{a}G$ is a right weak Artin-Rees ideal of R.

Taking $G = \langle 1 \rangle$ in 6.12 we obtain the following earlier result.

6.13 Corollary (Nouazé and Gabriel 1967) If R is a right Noetherian ring any polycentral ideal of R is right weak Artin-Rees.

6.14 Corollary. Let N be a nilpotent normal subgroup of the polycyclic-by-finite group G. If $\mathbf{a}$ is any ideal of $\mathbf{Z}N$ normalized by G, then $\mathbf{a}G$ is a right weak Artin-Rees ideal of $\mathbf{Z}G$.

For just take $R = \mathbf{Z}G$ and $S = \mathbf{Z}N$ in 6.12 and apply 6.5 with $J = \mathbf{Z}$. Actually we could get by with 6.13 and making no further use of 6.12. Now 6.13 is simpler to prove than 6.12, see Chatters and Hajarnavis (1980), p. 143. However we do now embark on a proof of 6.12.

Proof of 6.12 The proof copies the approach of that of 6.9. Let

$$\{0\} = \mathbf{a}_0 \le \mathbf{a}_1 \le \cdots \le \mathbf{a}_r = \mathbf{a}$$

be a polycentral series for the polycentral ideal $\mathbf{a}$. If a is a central element of S then $a^G = \{g^{-1}ag; \ g \in G\}$ is a central subset of S normalized by G. Thus we may assume that the $\mathbf{a}_i$ are all normalized by G. We work in the polynomial ring $R[X]$. Set

$$R^{\wedge} = \langle R, \mathbf{a}_i G X^{e(i)} : e(i) = 2^{r-i} \text{ for } i = 1, 2, \ldots, r \rangle.$$

6.12.1 If $R^{\wedge}$ is right Noetherian, then $\mathbf{a}G$ is right weak Artin-Rees.

Proof Here we make no use of polycentral so we can simplify notation by setting $G = \langle 1 \rangle$. Let N be a submodule of the finitely generated R-module M. Now

$$R^{\wedge} = R \oplus \mathbf{b}_1 X \oplus \mathbf{b}_2 X^2 \oplus \cdots \oplus \mathbf{b}_i X^i \oplus \cdots$$

for certain ideals $\mathbf{b}_i$ of R, where $\mathbf{b}_i \mathbf{b}_j \le \mathbf{b}_{i+j}$ for each i and j and $\mathbf{a}^i \le \mathbf{b}_i \le \mathbf{a}$. For example $\mathbf{b}_1 = \mathbf{a}$, $\mathbf{b}_2 = \mathbf{a}^2 + \mathbf{a}_{r-1}$, $\mathbf{b}_3 = \mathbf{a}^3 + \mathbf{a}_{r-1}\mathbf{a}$, $\mathbf{b}_4 = \mathbf{a}^4 + \mathbf{a}_{r-1}\mathbf{a}^2 + \mathbf{a}_{r-1}^2 + \mathbf{a}_{r-2}$, etc. Set

$$M^{\wedge} = M \oplus M\mathbf{b}_1 X \oplus M\mathbf{b}_2 X^2 \oplus \cdots \oplus M\mathbf{b}_i X^i \oplus \cdots.$$

Then again $M^{\wedge} = MR^{\wedge}$ is a finitely generated $R^{\wedge}$-module. Set

$$N^{\sim} = N \oplus (N \cap M\mathbf{b}_1) \oplus (N \cap M\mathbf{b}_2)X^2 \oplus \cdots \oplus (N \cap M\mathbf{b}_i)X^i \oplus \cdots.$$

Then $N^{\sim}$ is an $R^{\wedge}$-submodule of $M^{\wedge}$ and so is finitely generated. Consequently

$$N^{\sim} = (N \oplus (N \cap M\mathbf{b}_1)X \oplus (N \cap M\mathbf{b}_2)X^2 \oplus \cdots \oplus (N \cap M\mathbf{b}_k)X^k)R^{\wedge}$$

for some integer $k \geq 0$. Then equating coefficients of X^{k+1}, we have

$$N \cap M\mathbf{a}^{k+1} \leq N \cap M\mathbf{b}_{k+1} = N\mathbf{b}_{k+1} + (N \cap M\mathbf{b}_1)\mathbf{b}_k + \cdots + (N \cap M\mathbf{b}_k)\mathbf{b}_1 \leq M\mathbf{a},$$

since $\mathbf{a}^i \leq \mathbf{b}_i \leq \mathbf{a}$. □

6.12.2 The ring $R^\wedge$ is right Noetherian.

Proof Now $R^\wedge = \langle\langle S, \mathbf{a}_i X^{e(i)} : \text{for } e(i) = 2^{r-i} \text{ and } i = 1, 2, \ldots, r\rangle, G\rangle$. In view of P. Hall's version of the Hilbert Basis Theorem (see 3.7.3), it suffices to prove that

$$\langle S, \mathbf{a}_i X^{e(i)} : \text{for } e(i) = 2^{r-i} \text{ and } i = 1, 2, \ldots, r\rangle$$

is right Noetherian. Thus again we may assume that $G = \langle 1 \rangle$.

For simplicity we prove 6.12.2 only for $r = 2$. The general proof involves no more ideas, merely many more suffices. Thus we now have $\{0\} = \mathbf{a}_0 \leq \mathbf{a}_1 \leq \mathbf{a}$ and $R^\wedge = \langle R, \mathbf{a}_1 X^2, \mathbf{a}X \rangle$. Set $T = \langle R, \mathbf{a}_1 X, \mathbf{a}_1 X^2 \rangle$. Since $\mathbf{a}_1$ is centrally generated T is right Noetherian as in the proof of 6.9.

Let $z \in \mathbf{a}$ be central modulo $\mathbf{a}_1$ and let $t \in T$. Then t has the form

$$t = s + \sum a_1 a_2 \cdots a_u X^v$$

where $s \in R$, the $a_i \in \mathbf{a}_1$ and $u \leq v \leq 2u$. Then

$$[t, zX] = t(zX) - (zX)t$$
$$= [s, z]X + \sum \sum_i a_1 a_2 \cdots a_{i-1}[a_i, z]a_{i+1} \cdots a_u X^{v+1}.$$

Also

$$[a_i, z] \in [(\mathbf{a}_1 \cap \zeta_1(R))R, z]$$
$$\in [\mathbf{a}_1 \cap \zeta_1(R), z]R + \mathbf{a}_1[R, z] \subseteq \mathbf{a}_0 R + \mathbf{a}_1 \mathbf{a}_1 = \mathbf{a}_1^2.$$

Hence $[T, zX]$ is contained in $\mathbf{a}_1 X + \sum \mathbf{a}_1^{u+1} X^{v+1}$ and clearly $u + 1 \leq v + 1 \leq 2(u + 1)$. Thus $[T, zX] \subseteq T$. If $z' \in \mathbf{a}$ is also central modulo $\mathbf{a}_1$, then $[zX, z'X] \in \mathbf{a}_1 X \leq T$. Therefore $R^\wedge$ is right Noetherian by the following. □

6.12.3 Let $R = \langle S, Y \rangle$ be a ring with S a right Noetherian subring of R and Y a finite subset of R with $[S \cup Y, S \cup Y] \subseteq S$. Then R is right Noetherian.

Proof For let $y \in Y$ and set $Z = Y \backslash \{y\}$. By induction on $|Y|$ we may assume that $T = \langle S, Z \rangle$ is right Noetherian. Also $[S \cup Z, y] \subseteq S$, so $[T, y] \subseteq T$. Hence $Ty + T = T + yT$. Then R is Noetherian by 3.7.1. □

6.15 **Theorem** (Roseblade 1971) For any group G, the group ring $\mathbf{Z}G$ is a hyper-central ring if and only if G is a hypercentral group.

Half of this we have already; if G is hypercentral, then so is $\mathbf{Z}G$ by 6.5. For the reverse implication we need the following.

6.15.1 Let H be a normal subgroup of a group G and consider $\mathbf{Z}G$ with $\mathbf{g}$ and $\mathbf{h}$ the corresponding augmentation ideals as usual. Then as right G-modules we have

$$H/H' \cong_G (\mathbf{h}+\mathbf{gh})/\mathbf{gh} = (\mathbf{Z}+\mathbf{g})\mathbf{h}/\mathbf{gh} = \mathbf{Z}G\mathbf{h}/\mathbf{gh}.$$

Proof We prove first the case where $H = G$, namely that $H/H' \cong \mathbf{h}/\mathbf{h}^2$ as abelian groups. Now $\mathbf{Z}H = \bigoplus_{h \in H} \mathbf{Z}h$, so $\mathbf{h} = \bigoplus_{h \in H \setminus \langle 1 \rangle} \mathbf{Z}(h-1)$ is free abelian on the exhibited generators. Hence $(h-1) \mapsto hH'$ extends to a homomorphism $\sigma : \mathbf{h} \to H/H'$. If $h, k \in H$, then

$$(h-1)(k-1) = (hk-1) - (k-1) - (h-1) \tag{$*$}$$

so $(h-1)(k-1)\sigma = hkH'.k^{-1}H'.h^{-1}H' = H'$ and hence $\mathbf{h}^2\sigma = \langle 1 \rangle$. Thus σ factors as $\sigma' : \mathbf{h}/\mathbf{h}^2 \to H/H'$ through $\mathbf{h}/\mathbf{h}^2$. Conversely let $\tau : H \to \mathbf{h}/\mathbf{h}^2$ be the map $h \mapsto (h-1) + \mathbf{h}^2$. By $(*)$ the map τ is a homomorphism. Since $H\tau$ is abelian, $H'\tau = \langle 1 \rangle$ and τ factors as $\tau' : H/H' \to \mathbf{h}/\mathbf{h}^2$ through H/H'. Clearly σ' and τ' are inverse maps.

Now $\mathbf{gh} \cap \mathbf{h} = \mathbf{h}^2$; for trivially $\mathbf{h}^2 \le \mathbf{gh} \cap \mathbf{h}$ and if T is a transversal of H to G we have $th - 1 = (h-1) + (t-1)h$ for any h in H and t in T. Thus

$$\mathbf{gh} \le \mathbf{h}^2 + \sum_{t \in T \setminus H} (t-1)\mathbf{h}$$

and since $\mathbf{Z}G = \bigoplus_{t \in T} t\mathbf{Z}H$ we have $\mathbf{gh} \cap \mathbf{h} \le \mathbf{h}^2$.

Putting the above together we have

$$H/H' \cong_{\mathbf{Z}} \mathbf{h}/\mathbf{h}^2 = \mathbf{h}/(\mathbf{gh} \cap \mathbf{h}) \cong_{\mathbf{Z}} (\mathbf{h}+\mathbf{gh})/\mathbf{gh},$$

where the combined map ϕ of H/H' to $(\mathbf{h}+\mathbf{gh})/\mathbf{gh}$ is given by $hH' \mapsto (h-1) + \mathbf{gh}$. Finally if $h \in H$ and $g \in G$, then

$$(h-1)g = g(h^g - 1) = (h^g - 1) + (g-1)(h^g - 1),$$

so firstly $\mathbf{h}+\mathbf{gh}$ and $\mathbf{gh}$ are right G-submodules of $\mathbf{Z}G$, so $(\mathbf{h}+\mathbf{gh})/\mathbf{gh}$ is a right G-module, and secondly

$$(h^g H')\phi = (h^g - 1) + \mathbf{gh} = (h-1)g + \mathbf{gh} = ((h-1)+\mathbf{gh})g = (hH')\phi.g$$

and hence ϕ is a G-homomorphism. The proof of 6.15.1 is complete. $\qquad\square$

Proof of 6.15 Assume G is a group with $\mathbf{Z}G$ hypercentral. If N is a normal subgroup of G, then $\mathbf{Z}(G/N)$ is isomorphic to $\mathbf{Z}G/(N-1)\mathbf{Z}G$ and so is also hypercentral. Thus it suffices to prove that $\zeta_1(G) \neq \langle 1 \rangle$. Suppose G has a non-trivial abelian normal subgroup A. By 6.15.1 we have that A is isomorphic as right G-module to $(\mathbf{a} + \mathbf{ga})/\mathbf{ga}$. The latter contains a non-zero central element a. Then $ga = ag$ for any g in G. But $(\mathbf{a} + \mathbf{ga})/\mathbf{ga}$ is trivial as left G-module, so $ga = a$. Consequently $a \neq 0$ is a fixed point of $(\mathbf{a} + \mathbf{ga})\mathbf{ga}$ as right G-module. Hence A contains a non-trivial central element of G and so $\zeta_1(G) \neq \langle 1 \rangle$.

Now assume that G has no non-trivial abelian normal subgroups. Since $\mathbf{Z}G$ is hypercentral there exists some x in $\zeta_1(\mathbf{Z}G)\setminus\{0\}$, say $x = \alpha_1 g_1 + \alpha_2 g_2 + \cdots + \alpha_r g_r$, where the α_i are non-zero integers and the g_i are distinct elements of G. If $g \in G$, then

$$x = x^g = \alpha_1 g_1^g + \alpha_2 g_2^g + \cdots + \alpha_r g_r^g.$$

Thus the finite set $\{g_1, g_2, \ldots, g_r\}$ is permuted by G. Let $M = \langle g_1, g_2, \ldots, g_r \rangle$ and let $C = C_G(M)$. Then M is normal in G and the index $(G : C)$ is finite. Further $M \cap C$ is abelian and normal in G and consequently $M \cap C = \langle 1 \rangle$. Now $\mathbf{Z}(G/C) \cong \mathbf{Z}G/(C-1)\mathbf{Z}G$ is hypercentral and if G/C is nilpotent, then so is $M \cong MC/C \leq G/C$. But then G has a non-trivial abelian normal subgroup, for example $\zeta_1(M)$. Therefore it suffices to prove the result for G/C and consequently we may assume that G is finite.

Let $N = \gamma^\omega G = \bigcap_{1 \leq i < \omega} \gamma^i G$ and suppose p is a prime dividing $|N|$. Then $R = \mathbf{F}_p G$ is an image of $\mathbf{Z}G$ (with $\mathbf{F}_p$ being the field of p elements) and therefore is hypercentral. Let T be a transversal of N to G. Set $\sigma_N = \sum_{x \in N} x \in R$ and similarly define σ_T and σ_G. Then $\sigma_G = \sigma_N \sigma_T$. Since p divides $|N|$ we have that $\sigma_N = \sigma_N - |N| \in \mathbf{n}$, the augmentation ideal of N in $\mathbf{F}_p N$. Thus $\sigma_G \in \mathbf{n}G$; also σ_G is central in G.

Now G stabilizes the series $R \geq \mathbf{g} \geq \mathbf{g}^2 \geq \cdots \geq \mathbf{g}^i$ of right G-submodules, so $G/C_G((R/\mathbf{g}^i)_R)$ is nilpotent (see 1.19) and hence N centralizes $(R/\mathbf{g}^i)_R$ for every i. Thus

$$N \leq \bigcap_i G \cap (1 + \mathbf{g}^i) = G \cap (1 + \mathbf{g}^\omega)$$

and therefore $\sigma_G \in \mathbf{n}G \leq \mathbf{g}^\omega$.

By hypothesis $\mathbf{g}$ is polycentral as R is finite and therefore by 6.13 it is weak Artin-Rees. Hence there exists $n \geq 1$ with $(\sigma_G R) \cap \mathbf{g}^n \leq (\sigma_G R)\mathbf{g}$. That is, $\sigma_G R \cap \mathbf{g}^n = \{0\}$ for clearly $\sigma_G \mathbf{g} = \{0\}$. But $\sigma_G \in \mathbf{g}^\omega \leq \mathbf{g}^n$, so $\sigma_G = 0$, which is false. This contradiction completes the proof of the theorem. $\square$

6.16 Corollary (Roseblade 1971) For any group G, the group ring $\mathbf{Z}G$ is a polycentral ring if and only if G is a finitely generated nilpotent group.

Proof For if G is finitely generated and nilpotent, then $\mathbf{Z}G$ is hypercentral by 6.15 and is (left and right) Noetherian by 3.7. Consequently $\mathbf{Z}G$ is polycentral. Now assume that $\mathbf{Z}G$ is polycentral. Then G is hypercentral by 6.4. Thus G has an ascending central series

$$\langle 1 \rangle = G_0 \le G_1 \le \cdots \le G_\alpha \le \cdots \le G_\lambda = G$$

with each factor $G_{\alpha+1}/G_\alpha$ non-trivial and cyclic. Since the ideals of $\mathbf{Z}G$ are all finitely generated, $\mathbf{Z}G$ satisfies the maximal condition on ideals. Thus with $\mathbf{g}_\alpha = (G_\alpha - 1)\mathbf{Z}G_\alpha$ we have

$$\{0\} < \mathbf{g}_1\mathbf{Z}G \le \mathbf{g}_2\mathbf{Z}G \le \cdots \le \mathbf{g}_\alpha\mathbf{Z}G \le \cdots \le \mathbf{g}_\lambda\mathbf{Z}G = \mathbf{g}$$

contains only finitely many distinct members. But $(1 + \mathbf{g}_\alpha\mathbf{Z}G) \cap G = G_\alpha$ by 3.5.1. Thus λ is finite and G is finitely generated and nilpotent. $\qquad\square$

Above we have only touched on what is known in this area, but we have covered all we need. We conclude this chapter by just summarizing further related results. Throughout G will denote a polycyclic-by-finite group. We start with the Artin-Rees theorems. Jategaonkar (1974) effectively has the following. Let J be any ring and $\mathbf{a}$ an ideal of JG of finite index. If $\mathbf{a} \cap J$ contains a right strong Artin-Rees ideal $\mathbf{b}_1$ of finite index (in J), then JG contains a right strong Artin-Rees ideal $\mathbf{b}$ of finite index with $\mathbf{b}_1 \le \mathbf{b} \le \mathbf{a}$. Thus he deduces that every ideal of $\mathbf{Z}G$ of finite index contains a right strong Artin-Rees ideal of finite index.

Roseblade has a closely related result, which we now describe. If p is any prime, say G is p-nilpotent if every finite image of G is an extension of a p'-group by a p-group (recall we are assuming G is polycyclic-by-finite throughout). For any group H, let $\psi_p(H)$ denote the intersection of the centralizers of the finite chief factors of H of order divisible by p.

Exercise If H is polycyclic-by-finite prove that

(a) $\psi_p(H)$ has finite index in H and
(b) if N is a normal subgroup of H, then N is p-nilpotent if and only if $N \le \psi_p(H)$.

Then Roseblade (1976) and Segal (1975a) have the following. Let J be a right Noetherian ring, e a positive integer and H a normal subgroup of G that is p-nilpotent for every prime p dividing e. Then $\mathbf{b} = eJG + \mathbf{h}G$ is a right weak Artin-Rees ideal of JG. In this theorem suppose $J = \mathbf{Z}$ and $\mathbf{a}$ is an ideal of $\mathbf{Z}G$ of finite index. Set $e = |\mathbf{Z}G/\mathbf{a}|$ and $H = (1 + \mathbf{a}) \cap \bigcap_{p|e} \psi_p(G)$. Then $\mathbf{b} \le \mathbf{a}$, $\mathbf{b}$ is right weak Artin-Rees and $\mathbf{Z}G/\mathbf{b}$ is finite. Compare this with Jategaonkar's corollary above.

A further related result (see Roseblade and Smith (1979) for the field case and 4.3.9 of Shirvani and Wehrfritz (1986) in general) is the following. Suppose G is actually polycyclic and let J be a commutative Noetherian ring and $\mathbf{a}$ an ideal of JG of finite index. Suppose G is p-nilpotent for every prime p dividing the characteristic of $JG/\mathbf{a}$. Then $\mathbf{a}$ is right weak Artin-Rees.

Many more 'hypercentral' theorems are known than those we have proved above. Let J be any ring and H any group. If J and H are hypercentral, then so is JH

by 6.5. If JH is hypercentral, then so is J since $J \cong JH/\mathbf{h}$. However we cannot in general deduce that H is hypercentral, though 6.15 does say we can if $J = \mathbf{Z}$. It is very easy to check that FH is hypercentral for any finite group H if F is a field of characteristic zero or of characteristic prime to the order of H. More generally we have the following, see Roseblade and Smith (1976). Let F be a field of characteristic $p \geq 0$ and let H be any group. Then FH is hypercentral if and only if either $p = 0$ and H is hyper (central or finite), or $p > 0$ and H has an ascending normal series each factor of which is a finite p'-group or has its centralizer in H of finite index a power of p. They deduce that FH is polycentral if and only if either $p = 0$ and H is finitely generated and finite-by-nilpotent, or, $p > 0$ and H is finitely generated and a finite p'-group by nilpotent by a finite p-group.

We return to consideration of our polycyclic-by-finite group G. In a series of papers between 1971 and 1979, written singularly and jointly, Roseblade and P. Smith produce the following four results (some of whose implications follow from what we have proved above). Let F be a field of characteristic $p \geq 0$ and N a normal subgroup of G. If $p = 0$ the following are equivalent: (a) $\mathbf{n}$ is polycentral in FN; (b) $\mathbf{n}G$ is right weak Artin-Rees in FG; (c) N is finite by nilpotent. If $p > 0$ the following are equivalent: (a) $\mathbf{n}G$ is right weak Artin-Rees in FG; (b) N is p-nilpotent.

Keeping this notation, but taking $N = G$, they deduce the following. If $p = 0$ the following are equivalent: (a) $\mathbf{g}$ is polycentral (in FG); (b) FG is polycentral; (c) $\mathbf{g}$ is right weak Artin-Rees; (d) FG is a weak Artin-Rees ring; (e) G is finite-by-nilpotent. If $p > 0$, the following are equivalent: (a) $\mathbf{g}$ is right weak Artin-Rees; (b) FG is a weak Artin-Rees ring; (c) G is p-nilpotent. If H is a polycyclic-by-finite group that is not nilpotent-by-finite, set $G = \psi_p(H)$. Then G is p-nilpotent but not a finite p'-group by nilpotent by a finite p-group and hence FG is not polycentral. Thus we cannot add polycentrality to our list of equivalent conditions when $p > 0$ as we can when $p = 0$.

Our 6.15 is the integral analogue of some of what we have summarized above. We conclude this chapter by listing the integral analogues of some of the remainder. Let N be a normal subgroup of our polycyclic-by-finite group G. Then (Roseblade and Smith) the following are equivalent: (a) $\mathbf{n}$ is polycentral in $\mathbf{Z}N$; (b) $\mathbf{n}G$ is right weak Artin-Rees in G; (c) N is nilpotent. The following are equivalent: (a) $\mathbf{g}$ is polycentral; (b) $\mathbf{Z}G$ is polycentral; (c) $\mathbf{g}$ is right weak Artin-Rees; (d) $\mathbf{Z}G$ is a weak Artin-Rees ring; (e) G is nilpotent.

Chapter 7
Groups Acting on Finitely Generated Commutative Rings

Let G be a polycyclic group, $\mathbf{b}$ an ideal of the group ring $\mathbf{Z}G$ and A an abelian normal subgroup of G. Put $R = \mathbf{Z}G/\mathbf{b}$ and let S denote the subring of R generated by the image of A. Then S is a finitely generated commutative ring and G acts on S by conjugation and normalizes the image of A. We wish to work by induction. It is not sufficient to know about the group rings $\mathbf{Z}(G/A) \cong \mathbf{Z}G/(A-1)\mathbf{Z}G$ of G/A and $\mathbf{Z}A$ of A, say by induction on the Hirsch number. We also need to allow for how G acts on $\mathbf{Z}A$ and more generally on $\mathbf{S}$. In this chapter we do the basic groundwork for this.

Let $F \leq K$ be fields and A an *ordered abelian group* (that is an abelian group with a total order preserved by addition). A *valuation* v of K over F with *value group* A is a map v of K onto $A \cup \{\infty\}$ such that

(a) $v(x) = \infty$ if and only if $x = 0$;
(b) $v(xy) = v(x) + v(y)$ for all x and y in K;
(c) $v(x+y) \geq \min\{v(x), v(y)\}$ for all x and y in K;
(d) $v(x) = 0$ for all $x \in F^* = F\backslash\{0\}$.

Part (b) essentially says that v is a homomorphism of K^* onto A. In part (c) we have equality if $v(x) \neq v(y)$.

7.1 If A is an ordered abelian group, then A is torsion-free. Thus if F is a field, then FA is a domain and therefore has a quotient field K. There is a valuation v of K over F with value group A. For if $x = \alpha_1 a_1 + \alpha_2 a_2 + \cdots + \alpha_r a_r \in FA$, where the α_i lie in F^* and the a_i lie in A with $a_1 < a_2 < \cdots < a_r$, set $v(x) = a_1$. If $f/g \in K$ with f and g in FA, set $v(f/k) = v(f) - v(g)$. Now check that everything works.

7.2 Let $F \leq K \leq L$ be fields, with L a finite extension of K. Let v be a valuation of K over F. Then there exists a valuation w of L over F such that $w|_K = v$. Also $v(K^*)$ has finite index in $w(L^*)$.

See Knight (1971), 7.2.2 and 7.3.3.1 or Cohn (1974), 2nd ed. Vol. 2, 8.5.1 and 8.5.3, or just about any book on valuation theory.

B.A.F. Wehrfritz, *Group and Ring Theoretic Properties of Polycyclic Groups*, Algebra and Applications 10,
DOI 10.1007/978-1-84882-941-1_7, © Springer-Verlag London Limited 2009

The following theorem is critical for much of the rest of this book. It is due to G.M. Bergman, though the version below is due to Roseblade (1971).

7.3   Theorem (Bergman 1971) Let R be an integral domain generated by its sub-field F and the finitely generated subgroup U of its group of units (so $R = F[U]$). For each valuation v over F of the quotient field K of R, set $U_v = \{u \in U : v(u) = 0\}$ and denote the transcendence degree of K over F by d. Then the set

$$\{U_v : \text{rank}(U/U_v) = d\}$$

of subgroups of U is finite and non-empty.

Note that since the value group A_v of v is torsion-free and $v|_U : U \to A_v$ is a homomorphism, U_v is a subgroup of U and U/U_v is a free abelian group of finite rank. If A is any finitely generated abelian group, by $\text{rank}(A)$ in this chapter we mean the torsion-free rank of A, namely the rank of the free abelian group $A/\tau(A)$.

Proof There exists a torsion-free subgroup W of U such that W is linearly inde-pendent over F (that is, $F[W] = FW$, the group algebra) and R is algebraic over $F[W]$. Necessarily $\text{rank}(W) = d$. Make W into an ordered group (e.g. by embed-ding W into the additive group of the reals $\mathbf{R}$). By 7.1 there is a valuation of $F(W)$ over F with value group W inducing a bijection on W. Now K is algebraic and finitely generated over $F(W)$, so by 7.2 we can extend this valuation to a valuation v of K. Moreover $v(K^*)$ is torsion-free abelian and a finite extension of W. Thus $v(K^*) \cong W$. Also $W \le U$, $W \cap U_v = \langle 1 \rangle$ and U/U_v embeds into $v(K^*)$. Therefore $\text{rank}(U/U_v) = \text{rank}(W) = d$. Let $\mathcal{V}$ denote the set of all valuations v of K over F such that $\text{rank}(U/U_v) = d$. The above shows that $\mathcal{V}$ is not empty.

Let $\mathcal{S}$ denote the set of all finite sets S of isolated subgroups of U such that every member of $\mathcal{V}$ kills some member of S and no member of S contains any other member of S. Since U_v is isolated in U for any v, the set whose only member is the torsion subgroup of U lies in $\mathcal{S}$, so $\mathcal{S}$ is not empty.

If S and T belong to $\mathcal{S}$ write $S \le T$ if every member of T contains some member of S and denote by $S \vee T$ the set of minimal elements under containment of the set of isolators in U of subgroups of U of the form PQ for P in S and Q in T. Since each U_v is isolated it follows that $S \vee T$ lies in $\mathcal{S}$. It is elementary that $\le$ is a partial order on $\mathcal{S}$ with $S, T \le S \vee T$.

We claim that ascending chains in $\mathcal{S}$ have finite length. Let S and T be elements of $\mathcal{S}$ with $S < T$ (meaning of course $S \le T$ and $S \ne T$). Let s_i (resp. t_i) be the number of elements of S (resp. T) of torsion-free rank $i \ge 0$. If j is the least rank of a member of $S \backslash T$, and such j does exist since $S < T$, then $s_i = t_i$ for $i < j$ and $s_j > t_j$: for if $P \in S$ has rank less than j, then $P \in T$ by the choice of j and if $Q \in T$ has rank $i \le j$ then Q contains an element P of S by definition of the order $\le$ and either $\text{rank}(P) < j$, whence $P \in T$ and $P = Q$, or $\text{rank}(P) = j = \text{rank}(Q)$ and $P = Q$ as P is isolated. Consequently ordering lexicographically

$$(t_0, t_1, \ldots, t_{\text{rank}(U)}) < (s_0, s_1, \ldots, s_{\text{rank}(U)})$$

and the claim follows. Thus S contains a maximal member M say. This M is unique since $M \leq M \vee S$ for any $\mathbf{S}$ in S.

Let $w \in V$ and choose $P \in M$ with $w(P) = \{0\}$. It suffices to prove that $P = U_w$, for then $\{U_v : v \in V\} \subseteq M$ is finite. Assume otherwise. Since P is isolated there exists a complement Q of P in U. Clearly $P < U_w$, so

$$\operatorname{rank}(Q) > \operatorname{rank}(U/U_w) = d.$$

Thus Q is not linearly independent over F, say $\alpha_1 q_1 + \alpha_2 q_2 + \cdots + \alpha_n q_n = 0$, where the $\alpha_i \in F^*$ and the q_i are distinct elements of Q. If $v \in V$ then for some $i \neq j$ we have $v(\alpha_i q_i) = v(\alpha_j q_j)$; otherwise by property (c) of valuations $\infty = v(0) = v(q_i) \in v(K^*)$ for some i, which is false. Thus for some $i \neq j$ we have $v(q_i q_j^{-1}) = 0$.

Let S be the set of minimal elements of the set of isolators in U of all subgroups of U of the form $\langle q_i q_j^{-1} \rangle$ for $i, j = 1, 2, \ldots, n$ with $i \neq j$. The above shows that $S \in S$ and hence we have $S \subseteq M$. But then P contains $q_i q_j^{-1}$ for some $i \neq j$, which contradicts the definition of Q. This contradiction proves that $P = U_w$ and completes the proof of the theorem. $\qquad\square$

7.4 Corollary (Bergman 1971) Let $R = F[U]$ be an integral domain generated by the field F and the finitely generated subgroup U of its group of units. Let G be a group of automorphisms of R normalizing F and U. Let W_0 be a torsion-free subgroup of U that is linearly independent over F. Let d be the transcendence degree of R over F. Then there exists a valuation v of the quotient field of R over F such that $V = \{u \in U : v(u) = 0\}$ satisfies

(a) the index $(G : N_G(V))$ is finite;
(b) $F \cap U \subseteq V$;
(c) $W_0 \cap V = \langle 1 \rangle$;
(d) $\operatorname{rank}(U/V) = d$.

Proof Let v be the valuation constructed in the first paragraph of the proof of 7.3 with F, R and U as given and W chosen with $W_0 \leq W \leq U$. Clearly (b), (c) and (d) hold. Also G permutes the members of V and hence G permutes the finite set $\{U_w : w \in V\}$. But $V = U_v$, so (a) follows. $\qquad\square$

Let $R \neq \{0\}$ be a commutative Noetherian ring. If R contains an ideal $\mathbf{n} \neq R$ that does not contain a product of (not necessarily distinct) prime ideals, then it has a maximal such ideal $\mathbf{m}$. Now $\mathbf{m}$ clearly cannot be prime so there exist ideals $\mathbf{a}$ and $\mathbf{b}$ of R properly containing $\mathbf{m}$ with $\mathbf{ab}$ contained in $\mathbf{m}$. By the choice of $\mathbf{m}$ both $\mathbf{a}$ and $\mathbf{b}$ contain products of prime ideals of R. Consequently so does $\mathbf{m}$. This is a contradiction and hence no such ideals $\mathbf{n}$ exist.

In particular there are finitely many prime ideals $\mathbf{p}_1, \mathbf{p}_2, \ldots, \mathbf{p}_s$ of R whose product is $\{0\}$. Suppose that the $\mathbf{p}_i$ for $1 \leq i \leq r$ are exactly the distinct minimal members of the set $\{\mathbf{p}_1, \mathbf{p}_2, \ldots, \mathbf{p}_s\}$. Then $(\prod_{1 \leq i \leq r} \mathbf{p}_i)^s = \{0\}$. If $\mathbf{p}$ is any prime ideal of R, then $(\prod_{1 \leq i \leq r} \mathbf{p}_i)^s \leq \mathbf{p}$ and so $\mathbf{p}_i \leq \mathbf{p}$ for some $i \leq r$. It follows that $\mathbf{p}_1, \mathbf{p}_2, \ldots, \mathbf{p}_r$ are exactly the minimal prime ideals of R.

7.5   Corollary (Bergman 1971) Let U_1 be a free abelian group of finite rank that is a *plinth* for the group G (meaning that G and all its subgroups of finite index act rationally irreducibly on U_1). Suppose that F is a field and $\mathbf{a} \neq \{0\}$ is a G-invariant ideal of the group algebra FU_1. Then $\dim_F(FU_1/\mathbf{a})$ is finite.

Proof Let $\mathbf{p}$ be a minimal prime of FU_1 over $\mathbf{a}$, set $R = FU_1/\mathbf{p}$ and let U denote the image of U_1 in R. By the remark above there are only finitely many minimal primes over $\mathbf{a}$, so $N_G(\mathbf{p})$ has finite index in G. Choose V as in 7.4 with $N_G(\mathbf{p})$ for G. Since U_1 is a plinth for G, if U/V is infinite, then $V = \{0\}$ and $\mathrm{rank}(U) = \mathrm{rank}(U_1)$, which exceeds the transcendence degree d of R (recall $\mathbf{a} \neq \{0\}$). This contradiction shows that U/V is finite, $d = 0$ and R is algebraic over F. Thus $\dim_F R$ is finite. Now if $\mathbf{p}_1, \mathbf{p}_2, \ldots, \mathbf{p}_r$ are the minimal primes over $\mathbf{a}$, then $(\mathbf{p}_1 \mathbf{p}_2 \cdots \mathbf{p}_r)^s \leq \mathbf{a}$ for some integer s. By the above each $\dim_F(FU_1/\mathbf{p}_i)$ is finite. Therefore $\dim_F(FU_1/\mathbf{a})$ is finite.   □

The work of Bergman above extends from U a finitely generated abelian group to U an abelian group of finite rank, a fact that is used, for example, in Wehrfritz (1991b), see Sect. 2. For 7.2 holds with L an algebraic extension provided we weaken the conclusion on $v_1(L^*)/v(K^*)$ from finite to periodic.

7.3 holds with U just of finite rank and the proof needs very little modification. U/U_v is now just torsion-free of finite rank. Choose W as before, still free abelian. K is now just algebraic over $F(W)$, but the valuation v still exists. Here $v(K^*)$ is torsion-free abelian and a periodic extension of W. In general $v(K^*)$ and W need not be isomorphic, but they do have the same rank, $W \leq U$, $W \cap U_v = \langle 1 \rangle$ and U/U_v embeds into $v(K^*)$. Consequently we still have $\mathrm{rank}(U/U_v) = \mathrm{rank}(W) = d$. The construction of $M \in \mathcal{S}$, $P \in M$ and w proceeds exactly as before. Choose Q free abelian with $P \cap Q = \langle 1 \rangle$ and U/PQ periodic. Then $\mathrm{rank}(Q) = \mathrm{rank}(U/P) > \mathrm{rank}(U/U_w) = d$. The proof is now completed exactly as in that of 7.3.

7.4 remains true with U of finite rank and 7.5 with U_1 torsion-free of finite rank. The proofs require no modifications other than replacing 7.3 by its beefed up version sketched above.

If S is a subring of a commutative ring R, then $r \in R$ is said to be *integral* over S if it satisfies a monic polynomial over S. The *integral closure* of S in R is the set of all elements r of R that are integral over S. See Cohn (1974), 2nd ed. Vol. 2 Sect. 8.4 or Kaplansky (1970) Sect. 1.2, or Knight (1971), Sect. 5.1 or Zariski and Samuel (1958) Vol. 1 Sect. V.1 or just about any introductory text on commutative rings.

**7.6   **Let $F \leq K$ be fields. Suppose $S \geq F$ is a finitely generated F-subalgebra of K. Let L denote the quotient field of S in K and I the integral closure of S in K. Assume K is a finite extension of L. Certainly S is Noetherian. Then I is Noetherian (Zariski and Samuel 1958, Theorem V.9 on p. 267 of Vol. 1) and K is the quotient field of I (if $x \in K$ with x algebraic over L, then we have an equation

$s_0 x^n + s_1 x^{n-1} + \cdots + s_n = 0$, where the s_i lie in S and $s_0 \neq 0$. Then $(s_0 x)^n + s_1 (s_0 x)^{n-1} + \cdots + s_0^{n-1} s_n = 0$, so $s_0 x \in I$ and x lies in the quotient field of I). Then I is a Krull ring, which means that:

(a) $I_{\mathbf{p}}$ is a discrete valuation ring (that is a principal ideal domain with a unique non-zero prime ideal) for every minimal non-zero prime ideal $\mathbf{p}$ of I, see Kaplansky (1970), Theorem 103;
(b) $I = \cap I_{\mathbf{p}}$, where $\mathbf{p}$ is as in (a), see Kaplansky (1970), Theorem 104;
(c) if $x \in K$ then $x \in I_{\mathbf{p}}$ for all but a finite number of $\mathbf{p}$ as in (a), see Kaplansky (1970), Theorem 88 (and comment on his p. 76).

7.7 **Lemma** (Brewster 1976; Roseblade) Let R be an integral domain generated by its subfield F and a finitely generated subgroup U of its group of units and let G be a group of automorphisms of R normalizing F and U. If W is an isolated subgroup of U that is maximal subject to ($W \neq U$ and the index $(G : N_G(W))$ is finite), then one of the following holds.

(a) R is isomorphic as $S = F[W]$ algebra to the group ring $S(U/W)$ of U/W over S.
(b) $\text{rank}(U/W) = 1$ and R is algebraic over S.
(c) $\text{rank}(U/W) > 1$ and R is integral over S.

Proof Let K be a quotient field of R and L the quotient field of S in K. Suppose $\text{rank}(U/W)$ equals the transcendence degree d of K over L. Since in any case U splits over W and R is an image of $S(U/W)$, in this case (a) holds.

Now assume that $d \neq \text{rank}(U/W)$. By 7.4 (taking L for F) there exists an isolated subgroup V of U such that $(G : N_G(V))$ is finite, $L \cap U \leq V$ and $\text{rank}(U/V) = d$. Then $W \leq L \cap U \leq V$ and $\text{rank}(U/W) > d = \text{rank}(U/V)$, so $W < V$. By the maximal choice of W we have $U = V$ and $d = 0$. Thus R is algebraic over L and hence over S.

Suppose further that $\text{rank}(U/W) > 1$. It remains only to show that R is integral over S. Let I denote the integral closure of S in K. Apply 7.6. Since U is finitely generated the set $\mathbf{P}$ of minimal non-zero primes $\mathbf{p}$ of I with U not contained in $I_{\mathbf{p}}$ is finite by 7.6(c). If U is not contained in I then $\mathbf{P}$ is non-empty by 7.6(b). Let $\mathbf{p} \in \mathbf{P}$. Clearly $N_G(W)$ permutes the elements of $\mathbf{P}$, so $\mathbf{p}$ is normalized by a subgroup H of G of finite index. Then H normalizes X, the intersection of U with the group of units of $I_{\mathbf{p}}$. Also $W \leq X$ and $X \neq U$, since U does not lie in $I_{\mathbf{p}}$. Finally $\text{rank}(U/X) = 1$ by 7.6(a). This contradicts either the maximality of W or the assumption $\text{rank}(U/W) > 1$. Consequently $U \subseteq I$ and hence $R \leq I$ follows. $\square$

Note that in 7.7, case (b) really does arise in that R need not be integral over S. For example let w be an indeterminate over F and set $R = F[w, w^{-1}, u = (1 + w)^{-1}]$, $W = \langle w \rangle$, $U = \langle u, w \rangle$ and $G = \langle 1 \rangle$. Then U/W is infinite cyclic, but u is not integral over $F[W]$.

We now start to make some use of Zariski topologies, see the final few pages of Chap. 4. In particular we make use of groups acting connectedly, that is acting as a connected group of linear maps, on finite-dimensional vector spaces. The following corollary we make no use of here. It originally arose independently of 7.7 during a study of groups of semilinear maps, see Chap. 10.

7.8   Corollary (Wehrfritz 1977) Let R be an integral domain generated by its subfield F and a finitely generated subgroup U of its group of units and let G be a group of automorphisms of R normalizing F and U such that $\mathbf{Q} \otimes_{\mathbf{Z}} U$ is completely reducible as $\mathbf{Q}G$-module. Then there exists a torsion-free subgroup V of U with V a direct sum of direct summands of U such that:

(a) the index $(G : N_G(V))$ is finite;
(b) V is linearly independent over F, so $S = F[V]$ is just the group algebra of V over F;
(c) R is integral over S.

Proof U contains a torsion-free direct summand U_0 of finite index that is normalized by a normal subgroup N of G of finite index. Also R is integral over $F[U_0]$ and by Clifford's Theorem U_0 is rationally completely reducible as N-module. Hence we may assume that U is torsion-free and also that G acts connectedly on $QU = \mathbf{Q} \otimes_{\mathbf{Z}} U$.

Suppose first that G centralizes U and let $u_1, u_2, \ldots, u_n$ be a free basis of U. If U is linearly independent over F set $V = U$. Otherwise there is a non-zero polynomial $f(X_1, X_2, \ldots, X_n)$ over F with $f(u_1, u_2, \ldots, u_n) = 0$. If

$$\{X_1^{e(i,1)} X_2^{e(i,2)} \cdots X_n^{e(i,n)} : i \in I\}$$

is the set of monomials appearing in f set $N = 1 + \max\{e(i, j)\}$ and $r(j) = N^{j-1}$. Then

$$g(X_1, X_2, \ldots, X_n) = f(X_1, X_2 X_1^{r(2)}, \ldots, X_n X_1^{r(n)})$$

has a unique monomial of highest degree in X_1 and a unique monomial of lowest degree in X_1, since the $\sum_j e(i, j) r(j)$ for i in I are all distinct. For $j > 1$ set $z_j = u_j u_1^{-r(j)}$ and set $Z = \langle z_2, z_3, \ldots, z_n \rangle \leq U$. Then $g(u_1, z_2, \ldots, z_n) = 0$ and u_1 and u_1^{-1} are integral over $F[Z]$. Note that $U = \langle u_1 \rangle \times Z$. Induction on n now completes the proof in this case.

We now turn to the general case. If every irreducible $\mathbf{Q}G$-constituent of $\mathbf{Q} \otimes_{\mathbf{Z}} U$ is one-dimensional G centralizes U. The desired conclusion then follows from the case above, so suppose otherwise. Since QU is completely $\mathbf{Q}G$-reducible, there is a maximal isolated G-invariant subgroup W of U with $\text{rank}(U/W) > 1$. By induction on $\text{rank}(U)$ choose a G-invariant direct sum $V_1 \leq W$ of direct summands of W such that V_1 is linearly independent over F and $F[W]$ is integral over $F[V_1]$. Again by the complete reducibility hypothesis there exists a G-invariant isolated subgroup W_1 of U with $W_1 \cap W = \langle 1 \rangle$ and $U/W_1 W$ finite. By 7.7 either R is integral over $F[W]$ or W_1 is linearly independent over $F[W]$ and necessarily W_1 is a direct summand of U. In the first case set $V = V_1$ and in the second set $V = V_1 W_1$.   □

7.9   **Theorem** (Roseblade 1978) Let $R = \langle J, U \rangle = J[U]$ be an integral domain generated by its subring J and the finitely generated subgroup U of its group of units. Let G be a group of J-automorphisms (as a ring) of R normalizing U. Set $\Delta = \Delta_U(G) = \{u \in U : (G : C_G(u)$ is finite$\}$. Then some and hence any transversal of Δ to U is linearly independent over $S = J[\Delta]$.

Exercise In 7.9 prove that Δ is a subgroup of U with U/Δ torsion-free.

Thus U splits over Δ, say $U = \Delta \times V$. Then 7.9 says that R is the group ring of V over S.

Proof Let F be the quotient field of J in a quotient field of R. Then G acts as a group of F-automorphisms on $F[U]$. If a transversal T of Δ to U is linearly independent over $F[\Delta]$ then it surely is over $J[\Delta]$. Therefore we may assume that $J = F$ is a field.

We can replace G by any one of its subgroups of finite index. Also some subgroup of G of finite index centralizes the torsion subgroup C of U and normalizes a complement of C in U. Further $F[C]$ is a field and $C \le \Delta$. Hence we may assume that U is torsion-free. Then U induces a Zariski topology on $G/C_G(U)$. Again replacing G by a subgroup of finite index we may assume that G acts connectedly on U. (This reduction is not strictly necessary; it merely avoids repeatedly replacing G by smaller and smaller subgroups of finite index. In particular since $(G : C_G(\Delta))$ is finite we have $\Delta = C_U(G)$. Also U/Δ is torsion-free.

Let K be a quotient field of R, L the quotient field of S in K and denote the transcendence degree of K over L by d. If $d = \text{rank}(U/\Delta)$ the conclusion holds, so assume otherwise and seek a contradiction. By 7.4 with L for F there exists a G-invariant (using $G/C_G(U)$ is connected) isolated subgroup V of U such that $\Delta \le L \cap U \le V$ and $\text{rank}(U/V) = d$. Since $d < \text{rank}(U/\Delta)$ by hypothesis, so $\Delta < V$. If $V = U$ then K is algebraic over L and G embeds into the Galois group of K over L, which is finite. In this case $G = \langle 1 \rangle$ and $\text{rank}(U/\Delta) = 0 = d$, which we have assumed otherwise. Therefore $V \ne U$.

Let W be an isolated G-invariant subgroup of U that is maximal subject to $\Delta \le W < U$. By the above such a W exists and for any such W we have $\Delta < W$. By induction on $\text{rank}(U/\Delta)$ we may assume that any transversal of Δ to W is linearly independent over S. If a complement of W in U is linearly independent over $F[W]$ then $d = \text{rank}(U/\Delta)$, which is false. Hence R is algebraic over $F[W]$ by 7.7. In particular $d = \text{rank}(W/\Delta)$.

Let W_1 be any isolated G-invariant subgroup of W with $\Delta \le W_1 < W$ and let D be a complement of W_1 in W. Note that such W_1 exist since $\Delta \ne W$ and that D is linearly independent over $F[W_1]$. Apply 7.4 with the quotient field of $F[W_1]$ for F and D for W. Thus there exists a valuation v of K over $F[W_1]$ such that $W_2 = \{u \in U : v(u) = 0\}$ is isolated and G-invariant with $W_1 \le W_2$, $D \cap W_2 = \langle 1 \rangle$ and

$$\text{rank}(U/W_2) = \text{transcendence degree}(R/F[W_1]) = \text{rank}(W/W_1),$$

where we have used that R is algebraic over $F[W]$ and that a transversal of Δ to W is linearly independent over S. Since $W = W_1 D$ it follows that $W \cap W_2 = W_1$ and that U/WW_2 is finite.

Suppose first that W_1 is maximal subject to being isolated, G-invariant and satisfying $\Delta \leq W_1 < W$. Then W_2 is a maximal isolated G-invariant subgroup of WW_2 and so of U. Also the valuation ring of v (that is, $\{x \in K : v(x) \geq 0\}$) contains W_2, does not contain D and is integrally closed (Kaplansky 1970, Theorem 50). Thus R is not integral over $F[W_2]$ and so rank$(U/W_2) = 1$ by 7.7. But rank$(W/\Delta) = d, = $ rank(W_2/Δ), since this holds for any such W. Therefore rank$(U/W) = 1$.

Suppose G does not centralize U/Δ. Then choose W so that W/Δ contains $C_{U/\Delta}(G)$ (possible since $C_{U/\Delta}(G)$ is isolated) and now set $W_1 = \Delta$. Then since rank$(U/W) = 1$, so W_2/Δ is infinite cyclic and $W_2/\Delta \leq C_{U/\Delta}(G)$. This contradicts $W \cap W_2 = \Delta$. Consequently G centralizes U/Δ.

Let T be a complement of Δ in U. By hypothesis T is not linearly independent over S, so let $\sum_{1 \leq i \leq n} \alpha_i t_i = 0$ be a linear dependence with the integer n minimal, where the α_i are non-zero elements of S and the t_i are distinct elements of T. By multiplying on the right by t_1^{-1} we may assume that $t_1 = 1$. Clearly $n > 1$, so $t_2 \notin \Delta$ and hence there exists g in G with $t_2^g \neq t_2$. Then

$$0 = \left(\sum_i \alpha_i t_i\right)^g - \left(\sum_i \alpha_i t_i\right) = \sum_{2 \leq i \leq n} \alpha_i([t_i, g] - 1)t_i,$$

since G centralizes S. Now S is an integral domain, each α_i is non-zero and each $[t_i, g] \in \Delta \subseteq S$. Thus the minimal choice of n yields $[t_i, g] = 1$ for $i \geq 2$, which by construction is false for $i = 2$. This final contradiction completes the proof of the theorem. $\square$

7.10 Corollary (Roseblade 1978) Let J be a commutative ring, U a finitely generated abelian group and G a group of automorphisms of U. If $\mathbf{p}$ is a prime ideal of JU such that $U \cap (1 + \mathbf{p})$ and $(G : N_G(\mathbf{p}))$ are finite, then for $\Delta = \Delta_U(G)$ we have $\mathbf{p} = (\mathbf{p} \cap J\Delta)JU$.

When $\mathbf{p} = (\mathbf{p} \cap J\Delta)JU$, Δ is said to *control* $\mathbf{p}$. If $\mathbf{p}$ is any prime ideal of JU with $(G : N_G(\mathbf{p}))$ finite, we can replace G by $N_G(\mathbf{p})$, factor out by $\mathbf{p}^+ = U \cap (1+\mathbf{p})$ and use $JU/(\mathbf{p}^+ - 1)JU \cong J(U/\mathbf{p}^+)$ to reduce to the case where $\mathbf{p}^+ = \langle 1\rangle$. Hence the following is immediate from 7.10.

7.11 Corollary (Roseblade 1978) Let J, U and G be as in 7.10. Let $\mathbf{p}$ be an almost G-invariant prime ideal of JU. Define D by $D/\mathbf{p}^+ = \Delta_{U/\mathbf{p}^+}(G)$. Then $\mathbf{p} = (\mathbf{p} \cap JD)JU$.

Proof of 7.10 Replacing G by a subgroup of finite index does not change $\Delta_U(G) = \Delta$. Hence we may assume that $\mathbf{p}$ is normalized by G. In particular G acts on

$U/\mathbf{p}^+$. Let $u\mathbf{p}^+ \in \Delta_{U/\mathbf{p}^+}(G)$. Then $\{u^g\mathbf{p}^+ : g \in G\}$ is finite, and so is $\mathbf{p}^+$. Thus $\bigcup_{g \in G} u^g\mathbf{p}^+$ is finite. Clearly it contains u^G, so u^G is finite and $u \in \Delta$. Trivially $\mathbf{p}^+ \le \Delta$ and $\Delta/\mathbf{p}^+ \le \Delta_{U/\mathbf{p}^+}(G)$. Consequently $\Delta_{U/\mathbf{p}^+}(G) = \Delta/\mathbf{p}^+$.

Apply 7.9 to the domain $JU/\mathbf{p} = \langle(J+\mathbf{p}/\mathbf{p}), U/\mathbf{p}^+\rangle$. Hence if T is a transversal of Δ to U then T is linearly independent over $J\Delta$ modulo $\mathbf{p}$. Let $x \in JU$. Then $x = \sum_{t \in T} \alpha_t t$, where the $\alpha_t \in J\Delta$ and almost all are zero. If $x \in \mathbf{p}$ the above shows that $\alpha_t \in \mathbf{p}$ for all t, so $\mathbf{p} \le (\mathbf{p} \cap J\Delta)JU$. The converse is clear. $\qquad\square$

Exercise In 7.8 assume also that G centralizes F. Show that the hypotheses can be weakened from $\mathbf{Q} \otimes_{\mathbf{Z}} U$ completely $\mathbf{Q}G$-reducible to $\mathbf{Q} \otimes_{\mathbf{Z}} U$ splits as $\mathbf{Q}G$-module over $\mathbf{Q} \otimes_{\mathbf{Z}} \Delta_U(G)$ and the conclusion strengthened to V is a direct sum of two direct summands of U.

7.12 Let F be a locally finite field, N a normal finitely generated subgroup of the group G and $\mathbf{a}$ an ideal of the group algebra $S = FN$. If $n = \dim_F(S/\mathbf{a})$ is finite, with $\mathbf{a}_G = \bigcap_{g \in G} \mathbf{a}^g$, we have $\dim_F(S/\mathbf{a}_G)$ is also finite.

Proof N acts on $S/\mathbf{a}$ by right multiplication and this induces a map of N into $\mathrm{Aut}_F(S/\mathbf{a}) \cong GL(n, F)$. Now $GL(n, F)$ is a locally finite group, since F is a locally finite field, and N is finitely generated, so the image of N in $GL(n, F)$ is finite. That is, $(N : N \cap (1 + \mathbf{a}))$ is finite. Since N has only a finite number of subgroups of a given finite index,

$$H = \bigcap_{g \in G}(N \cap (1 + \mathbf{a}))^g = N \cap (1 + \mathbf{a}_G)$$

has finite index in N. Then $S/\mathbf{a}_G$ is an image of $F(N/H)$ and therefore has finite dimension over F. $\qquad\square$

7.13 Let G be a finitely generated abelian subgroup of $GL(n, F)$ that is connected and irreducible, F some field. Then G contains a cyclic subgroup that is also connected and irreducible.

This is a version of a lemma of Passman (see Passman 1977, 12.3.1), namely that if G is a finitely generated abelian group, F a field and V an FG-module that is FH-irreducible for every subgroup H of G of finite index, then there exists x in G such that V is FH-irreducible for every subgroup H of $\langle x \rangle$ of finite index. For by Hilbert's Nullstellensatz (e.g. Kaplansky 1970, Sect. 1.3) $n = \dim_F V$ is finite so we may assume that $G \le GL(n, F)$ with $V = F^{(n)}$. By 7.13 we have $\langle x \rangle \le G^0$ connected and irreducible. If H is a subgroup of $\langle x \rangle$ of finite index and if U is an FH-submodule of V, since $N_{\langle x \rangle}(U)$ is closed in $\langle x \rangle$, so U is an $F\langle x \rangle$-submodule of V. Thus V must be FH-irreducible as required.

Proof of 7.13 Now $V = F^{(n)}$ is an irreducible FG-module, so $V \cong_{FG} FG/\mathbf{m} = K$, where $\mathbf{m}$ is a maximal ideal and K is a finite extension field of F. Also G acts faithfully on K, so regard G as a subgroup of K^*. Let $F_1, F_2, \ldots$ be all the subfields of K containing F except K itself. Set $G_i = F_i \cap G$. Then G_i is a subgroup of G and if T_i/G_i is the torsion subgroup of G/G_i, then each T_i/G_i is finite. Now F_i is a proper FG_i-submodule of K, so K is not FG_i-irreducible and $(G : G_i)$ is infinite, since $G \le GL(n, F)$ is connected and irreducible and G_i is closed. Thus $(G : T_i)$ is infinite.

If K is inseparable over F, suppose that F_i is the maximal separable subfield of K over F. Then K is purely inseparable over F_i and K^*/F_i^* is a p-group for $p = \text{char } F$. Thus G/G_i is a finitely generated abelian p-group, which contradicts $(G : G_i)$ infinite. Therefore K is separable over F. Consequently elementary Galois theory yields that there are only a finite number, say r, of the subfields F_i.

Since r is finite and each $(G : T_i)$ is infinite, 1.22 yields that $\bigcup_i T_i \ne G$. Pick $y \in G \setminus \bigcup T_i$ and set $\langle x \rangle = \langle y \rangle^0$. Certainly $\langle x \rangle \le GL(n, F)$ and is cyclic and connected. Also $x \in G \setminus \bigcup T_i$. Now $F\langle x \rangle + \mathbf{m}/\mathbf{m} \le FG/\mathbf{m} = K$ and is an F-subalgebra of K. It is therefore one of the F_i or K itself. By the choice of x it is K itself. Thus K is $F\langle x \rangle$-irreducible. That is, $\langle x \rangle \le GL(n, F)$ is irreducible. $\qquad\square$

7.14   Let A be a free abelian group of finite rank r, ξ an automorphism of A and χ the characteristic polynomial of ξ. If $n \in \mathbf{Z}$ is not an eigenvalue of ξ, then

$$(A : A(\xi - n)) = |\chi(n)|.$$

Note that if $n \in \mathbf{Z}$ is an eigenvalue of ξ then $n = \pm 1$: for 0 is not an eigenvalue of ξ since ξ is invertible and if n is an eigenvalue of ξ then n^{-1} is an eigenvalue of ξ^{-1} and consequently is integral over $\mathbf{Z}$. Thus $n = \pm 1$.

Proof Let $n \in Z$ and $\eta = \xi - n$. Since $\chi(X) = \det(X - \xi)$ we have

$$|\chi(n)| = |\det(\xi - n)| = |\det \eta|.$$

By the choice of n, η is one-to-one. We claim this is enough to yield that $(A : A\eta) = |\det \eta|$.

Now $A\eta \le A$ and $\eta : A \to A\eta$ is an isomorphism. There exist elements a_i of A and positive integers e_i with $A = \bigoplus_{1 \le i \le r} \langle a_i \rangle$ and $A\eta = \bigoplus_i \langle e_i a_i \rangle$. Set $b_i = (e_i a_i)\eta^{-1}$. Then $A = \bigoplus_i \langle b_i \rangle$. Let γ be the automorphism of A given by $a_i \mapsto b_i$ for each i. Then the matrix of $\gamma\eta$ with respect to the ordered basis $(a_1, a_2, \ldots, a_r)$ is $\text{diag}(e_1, e_2, \ldots, e_r)$. Therefore $\det \gamma\eta = e_1 e_2 \cdots e_r = (A : A\eta)$ clearly. Since $\det \gamma = \pm 1$, the lemma follows. $\qquad\square$

7.15   (Roseblade 1973a). Let A and ξ be as in 7.14 and suppose Z is any infinite subset of the integers $\mathbf{Z}$. Then $\bigcap_{n \in Z} A(\xi - n) = \{0\}$.

Proof Clearly we may assume that $|n| \neq 1$ for all $n \in Z$. Let B be the isolator in A of $\bigcap_{n \in Z} A(\xi - n)$. If $\chi(X) = X^r + \alpha_1 X^{r-1} + \cdots + \alpha_r$ is the characteristic polynomial of ξ on A, then $(A : A(\xi - n)) = |\chi(n)|$ for $n \in Z$ by 7.14. Also

$$|\chi(n)| \geq |n|^r - (|\alpha_1| + |\alpha_2| + \cdots + |\alpha_r|)|n|^{r-1},$$

which tends to infinity as $|n|$ tends to infinity. Therefore A/B is infinite. Also B is ξ-invariant.

If $n \in Z$, then n is not an eigenvalue of ξ on A/B and therefore $A(\xi - n) \cap B = B(\xi - n)$. Thus $\bigcap_{n \in Z} A(\xi - n) = \bigcap_{n \in Z} B(\xi - n)$ and the latter is $\{0\}$ by induction on r, the case where $r = 0$ being trivial. $\qquad \square$

7.16 **Lemma** (Roseblade 1973a) Let F be a finite field of characteristic p, A a free abelian group of finite rank and set $S = FA$. If ξ is any automorphism of A, the intersection of all the maximal ideals of S normalized by ξ is $\{0\}$.

Proof For $n \geq 1$ set $A_n = \langle a^\xi a^{-q} : a \in A \rangle = \text{Im}(\xi - q)$ for $q = |F|^n$. Now q is not an eigenvalue of ξ, so $\ker(\xi - q) = \{0\}$, $\text{rank}(A) = \text{rank}(A_n)$ and A/A_n is finite. Set $\mathbf{a}_n = (A_n - 1)S$. Then $S/\mathbf{a}_n \cong F(A/A_n)$ is a finite dimensional F-algebra, ξ raises each element of A/A_n to its q-th power and q is a power of $p = \text{char } F$. Thus ξ acts on $S/\mathbf{a}_n$ by raising each element of $S/\mathbf{a}_n$ to its q-th power. In particular each ideal of $S/\mathbf{a}_n$ is ξ-invariant. If J denotes the Jacobson radical of $S/\mathbf{a}_n$, then J is nilpotent. Hence for some large positive integer j we have that ξ^j maps J to $\{0\}$. But ξ is an automorphism, so $J = \{0\}$ and $\mathbf{a}_n$ is an intersection of ξ-invariant maximal ideals of S.

Let $\lambda = \sum_{a \in A} \lambda_a a \in S \setminus \{0\}$, where the λ_a lie in F and almost all are zero. Set $\Sigma = \{ab^{-1} : a \neq b$ and $\lambda_a \lambda_b \neq 0\}$. Clearly Σ is finite. If $\Sigma \cap A_n \neq \emptyset$ for all $n \geq 1$, then some σ in Σ lies in infinitely many of the A_n. Since $\sigma \neq 1$ by construction, this contradicts 7.15. Thus $\Sigma \cap A_n = \emptyset$ for some $n \geq 1$. Then the elements a with λ_a non-zero lie in a transversal of A_n to A and $\lambda \notin \mathbf{a}_n$. Consequently $\bigcap_n \mathbf{a}_n = \{0\}$ and the lemma is proved. $\qquad \square$

There are a number of generalizations of 7.16 known. Let S be a commutative ring generated as ring by the finitely generated subgroup U of its group of units and suppose Γ is a group of ring automorphisms of S normalizing U. For example in 7.16 we can take $S = FA$, $\Gamma = \langle \xi \rangle$ and $U = F^* A$, since the multiplicative group F^* of F is cyclic. Then the following four results hold, see Wehrfritz (1979a).

(a) If Γ is cyclic there exists a normal subgroup H of Γ of finite index such that the intersection of the maximal ideals of S normalized by H is $\{0\}$.
(b) If Γ is cyclic and S is a domain, often and possibly always the intersection of the maximal ideals of S normalized by Γ is $\{0\}$.
(c) 7.16 becomes false in general if we replace $\langle \xi \rangle$ by the free abelian group of rank 2.
(d) If S is a domain and Γ is finite with the intersection of the maximal ideals of S normalized by Γ zero, then Γ is cyclic.

There are also generalizations of 7.16 to situations beyond polycyclic, for example to where F is any finitely generated field and A is a torsion-free abelian minimax group, see Wehrfritz (1992).

7.17 Theorem (Roseblade 1973a) Let A be a free abelian normal subgroup of the polycyclic-by-finite group G and suppose A is a plinth for G, meaning that the connected component of $G/C_G(A)$ acts $\mathbf{Q}$-irreducibly on A. Suppose F is a locally finite field and $\mathbf{a}$ is a non-zero ideal of $S = FA$. Then there exists a maximal ideal $\mathbf{m}$ of S containing no G-conjugate of $\mathbf{a}$; that is, for no g in G does $\mathbf{m}$ contain $\mathbf{a}^g$.

Proof Let $\lambda = \sum_{a \in A} \lambda_a a \in S\setminus\{0\}$, where the λ_a lie in F. It suffices to find a maximal ideal $\mathbf{m}$ of S with $\lambda^G \cap \mathbf{m} = \emptyset$. Suppose first that F is finite.

Let $H/C_G(A)$ denote the connected component of $G/C_G(A)$ containing $\langle 1 \rangle$ in its action on A. Let T be a transversal of H to G and set $\lambda_0 = \prod_{t \in T} \lambda^t$; note that $\lambda_0 \neq 0$ since S is a domain. If $\mathbf{m}$ is a maximal ideal of S with $\lambda_0 H \cap \mathbf{m} = \emptyset$, then $\lambda^G \cap \mathbf{m} = \emptyset$. Thus we may assume that $G = H$. In particular $G/C_G(A)$ is abelian (use Clifford's Theorem and 4.13).

By 7.13 there exists x in G such that $\langle x \rangle$ acts connectedly and $\mathbf{Q}$-irreducibly on A (so A is a plinth for $\langle A, x \rangle$). By 7.16 the set $\mathbf{M}$ of maximal ideals of S normalized by x is infinite. Since $G/C_G(A)$ is abelian, $\mathbf{M}$ is normalized by G. Now for $\mathbf{m} \in \mathbf{M}$, the ring $S/\mathbf{m}$ has finite dimension over F by the Nullstellensatz, so by 7.12 we have $\dim_F(S/\mathbf{m}_G)$ finite. Thus the orbits of G on $\mathbf{M}$ are all finite.

Suppose $\lambda^G \cap \mathbf{m} \neq \emptyset$ for every $\mathbf{m}$ in $\mathbf{M}$. Since G has infinitely many orbits on $\mathbf{M}$ we have infinitely many members of $\mathbf{M}$ that contain λ. By definition of $\mathbf{M}$ each such $\mathbf{m} \in \mathbf{M}$ contains the ideal $\mathbf{b} = \lambda^{\langle x \rangle} S$. Apply 7.5 (Bergman's theorem) to $\mathbf{b}$. Then $S/\mathbf{b}$ has finite dimension over F and hence only finitely many members of $\mathbf{M}$ can contain $\mathbf{b}$. This contradiction shows that $\lambda^G \cap \mathbf{m} = \emptyset$ for at least one $\mathbf{m}$ in M. The proof of the case where F is finite is now complete.

Assume F is any locally finite field. Now $\lambda \in EA$ for some finite subfield E of F. By the above case there is a maximal ideal $\mathbf{n}$ of EA with $\lambda^G \cap \mathbf{n} = \emptyset$. Pick any maximal ideal $\mathbf{m}$ of S with $\mathbf{n} \leq \mathbf{m}$ (note that necessarily $F\mathbf{n} < FA$, so $\mathbf{m}$ does exist). Clearly $EA \cap \mathbf{m} = \mathbf{n}$. Consequently $\lambda^G \cap \mathbf{m} = \emptyset$ and the proof of the theorem is complete. $\square$

7.17 can be extended as follows, see Theorem 2 of Wehrfritz (1991a).

7.18 Theorem Let F be a field, A a free abelian group of finite rank, G a polycyclic-by-finite group acting on A and λ a non-zero element of the group algebra FA. Then there is a semiprime G-invariant ideal $\mathbf{a}$ of FA modulo which λ is a unit and A is finite.

In 7.18 if $\mathbf{b}$ is a non-zero ideal of FA and we pick some λ in $\mathbf{b}\setminus\{0\}$, then with $\mathbf{a}$ as above $\mathbf{b}$ is not contained in $\mathbf{a}$ and hence nor is any of its conjugates $\mathbf{b}^g$ for $g \in G$. Further with $\mathbf{m}$ a maximal ideal of FA containing $\mathbf{a}$ we have $\lambda^G \cap \mathbf{m} = \emptyset$, so $\mathbf{b}^g$ does not lie in $\mathbf{m}$ for every g in G. This extends 7.17.

Again with $\mathbf{a}$ as above, $FA/\mathbf{a}$ is finite dimensional over F and is semisimple. It therefore has only finitely many maximal ideals and these are permuted by G. If $\mathbf{m}$ is a maximal ideal of FA containing $\mathbf{a}$, then $(G : N_G(\mathbf{m}))$ is finite. Also λ is a unit modulo $\mathbf{m}$, so $\lambda \notin \mathbf{m}$ for any such $\mathbf{m}$ and hence $\lambda^G \cap \mathbf{m} = \emptyset$. If A is also a plinth for G, this becomes Theorem C of Harper (1980).

Apart from the extension above of 7.17 and Harper's Theorem, 7.18 can also be used to prove the following theorem of Lichtman (1992), again see Wehrfritz (1991a).

Let F be a field, G a polycyclic-by-finite group, $D = F(G)$, the Artinian ring of quotients of the group algebra FG (which does exist) and n a positive integer. Then the matrix ring $D^{n \times n}$ is locally super-residually finite-dimensional over F; specifically for every finitely generated F-subalgebra J of $D^{n \times n}$ and every finite subset X of J there exists an F-algebra homomorphism ϕ of J with $\dim_F J\phi$ finite such that ϕ is one-to-one on X and $(G \cap J)\phi$ is finite.

There are more results of this general type in Shirvani and Wehrfritz (1986), Chap. 4, especially Sect. 4.2. Also a number of applications are included there.

Segal (2000) considers a somewhat similar situation to that of 7.18, but with F a finitely generated commutative ring, A a reduced abelian minimax group and G just a soluble group. He is particularly interested here in certain types of G-invariant prime ideal of FA. Segal in (2001) and (2006) again considers prime ideals of FA, but now where F is a field and A is just abelian and minimax.

Chapter 8
Prime Ideals in Polycyclic Group Rings

If G is a polycyclic group and $\mathbf{p}$ is a prime ideal of the group ring $\mathbf{Z}G$, how big is $\mathbf{p}$? For example, is it determined by a relatively small subgroup of G? How long can a chain of prime ideals of $\mathbf{Z}G$ be? These are the sort of questions we consider in this chapter. The proofs frequently use induction on the Hirsch number, so we begin by looking at the connection between the prime ideals of $\mathbf{Z}G$ and the prime ideals of $\mathbf{Z}H$ for H a normal subgroup of G.

Let J be a ring, G a group and H a normal subgroup of G. A G-*ideal* of JH is an ideal of JH normalized by G. Call a G-ideal $\mathbf{a}$ of JH a G-*prime* if whenever $\mathbf{b}$ and $\mathbf{c}$ are G-ideals of JH such that $\mathbf{bc} \le \mathbf{a}$, then either $\mathbf{b}$ or $\mathbf{c}$ is contained in $\mathbf{a}$.

8.1 Lemma Let J be a commutative ring, G a group and H a normal subgroup of G.

(a) If $\mathbf{p}$ is a prime ideal of JG, then $\mathbf{p} \cap JH$ is a G-prime of JH.
(b) If $\mathbf{a}$ is a G-prime of JH and if G/H is an ordered group (for example, if G/H is infinite cyclic), then $\mathbf{p} = \mathbf{a}JG$ is a prime ideal of JG.
(c) If $\mathbf{a}$ is a G-prime of JH and if $JH/\mathbf{a}$ is a right Noetherian ring, then $\mathbf{a} = \mathbf{q}_1 \cap \mathbf{q}_2 \cap \cdots \cap \mathbf{q}_n$, where the $\mathbf{q}_i$ are distinct and are the minimal primes of JH over $\mathbf{a}$. Further G permutes the $\mathbf{q}_i$ transitively.

By an ordered group, we mean a group with a total order preserved by left and right multiplication. If R is a right Noetherian ring, then R has only a finite number of minimal prime ideals (for example see Chatters and Hajarnavis 1980, 1.16). Thus in particular the n in part (c)) of 8.1 is necessarily finite.

Proof (a) Let $\mathbf{b}$ and $\mathbf{c}$ be G-ideals of JH with $\mathbf{bc} \le \mathbf{p} \cap JH$. Since $\mathbf{b}$ and $\mathbf{c}$ are normalized by G, we have that $JG\mathbf{b} = G\mathbf{b}$ and $\mathbf{c}G$ are ideals of JG whose product lies in $\mathbf{p}$. Thus either $G\mathbf{b} \le \mathbf{p}$ or $\mathbf{c}G \le \mathbf{p}$. Therefore $\mathbf{p} \cap JH$ is a G-prime of JH.

(b) Let $\mathbf{b}$ and $\mathbf{c}$ be ideals of JG containing $\mathbf{p}$ with $\mathbf{bc} \le \mathbf{p}$. Let T be a transversal of H to G, ordered by some (fixed) ordering of G/H as an ordered group. If $\alpha \in JG$, then $\alpha = \sum_T \alpha_t t$, where each α_t lies in JH with almost all α_t zero and,

B.A.F. Wehrfritz, *Group and Ring Theoretic Properties of Polycyclic Groups*, Algebra and Applications 10, DOI 10.1007/978-1-84882-941-1_8, © Springer-Verlag London Limited 2009

given α, this expression is unique. If α is non-zero, let $\alpha^\sim$ be the first non-zero co-efficient of α in the chosen ordering of T and set $0^\sim = 0$. Since conjugation leaves the order on G/H invariant $\mathbf{b}^\sim = \{\alpha^\sim : \alpha \in \mathbf{b}\}$ and $\mathbf{c}^\sim$ are G-ideals of JH. Also for β and γ in JG we have $\beta^\sim \gamma^\sim = (\beta\gamma)^\sim k$ for some k in H (if $\beta = \beta^\sim t_1 + \cdots$ and $\gamma = \gamma^\sim t_2 + \cdots$ then $\beta\gamma = \alpha^\sim \beta^\sim t_1 t_2 + \cdots$ and $t_1 t_2 = kt$ for some k in H and t in T), so $\mathbf{b}^\sim \mathbf{c}^\sim \leq \mathbf{a}$ by the definition of $\mathbf{p}$. If $\mathbf{b}^\sim \leq \mathbf{a}$ but $\mathbf{b} \neq \mathbf{p}$, choose β in $\mathbf{b} \backslash \mathbf{p}$ of minimal length (the number of non-zero β_t). For some t in T the length of $\beta_1 = \beta - \beta^\sim t$ is less than that of β and $\beta^\sim t \in \mathbf{p} \leq \mathbf{b}$. The minimal choice of β yields that $\beta_1 \in \mathbf{p}$. But then $\beta \in \mathbf{p}$, which is false. Consequently $\mathbf{b}^\sim \leq \mathbf{a}$ implies that $\mathbf{b} = \mathbf{p}$. In the same way $\mathbf{c}^\sim \leq \mathbf{a}$ implies that $\mathbf{c} = \mathbf{p}$. Part (b) of 8.1 follows.

(c) Clearly G permutes the $\mathbf{q}_i$, so $\mathbf{n} = \bigcap_i \mathbf{q}_i$ is a G-ideal of JH. By definition $\mathbf{n}/\mathbf{a}$ is the prime radical of the right Noetherian ring $JH/\mathbf{a}$ and as such is nilpotent (each ideal of $JH/\mathbf{a}$ contains a product of prime ideals) so $\mathbf{n}^s \leq \mathbf{a}$ for some positive integer s. But $\mathbf{a}$ is a G-prime. Thus $\mathbf{n} = \mathbf{a}$, which proves the first claim. If say $\mathbf{q}_1, \mathbf{q}_2, \ldots, \mathbf{q}_r$ is a G-orbit with $r < n$, then $\mathbf{b} = \bigcap_{i \leq r} \mathbf{q}_i$ and $\mathbf{c} = \bigcap_{i > r} \mathbf{q}_i$ are G-ideals of JH with $\mathbf{bc} \leq \mathbf{a}$. Thus $\mathbf{b} = \mathbf{a}$ or $\mathbf{c} = \mathbf{a}$. If say $\mathbf{b} = \mathbf{a}$ we have $\mathbf{q}_n \geq \mathbf{a} = \mathbf{b} \geq \mathbf{q}_1 \mathbf{q}_2 \cdots \mathbf{q}_r$. Then primeness yields $\mathbf{q}_n \geq \mathbf{q}_i$ for some $i \leq r < n$; whence $\mathbf{q}_n = \mathbf{q}_i$ by minimality of the $\mathbf{q}_j$, which is false. Similarly we obtain a contradiction if $\mathbf{c} = \mathbf{a}$. Consequently $r = n$. $\qquad\square$

We now extend 7.10 slightly. Let H be a finitely generated FC-group, say $H = \langle h_1, h_2, \ldots, h_n \rangle$. (An FC-group is one in which every element has only a finite number of conjugates.) Then by hypothesis each index $(H : C_H(h_i))$ is finite and clearly $\bigcap_i C_H(h_i) = Z$, the centre of G. Thus $(H : Z)$ is finite. By a theorem of Schur (see 1.18), the derived subgroup H' is also finite.

8.2 Corollary (Roseblade 1978) Let J be a commutative ring, H a finitely generated FC-group and G a group of automorphisms of H. If $\mathbf{p}$ is a prime ideal of JH such that $H \cap (1 + \mathbf{p})$ and $(G : N_G(\mathbf{p}))$ are finite, then for

$$\Delta = \Delta_H(G) = \{h \in H : (G : C_G(h)) < \infty\}$$

we have $\mathbf{p} = (\mathbf{p} \cap J\Delta)JH$.

Proof By the remarks above H contains a finite normal subgroup K such that H/K is torsion-free abelian. Clearly $K \leq \Delta$. Also $\Delta/K = \Delta_{H/K}(G)$. In particular H/Δ is torsion-free abelian ($\Delta_{H/K}(G)$ is the centralizer in H/K of the connected component of G in its action on H/K) and so is orderable (just embed it in the additive group of the reals). Set $\mathbf{q} = (\mathbf{p} \cap J\Delta)JH$. By 8.1(a) and 8.1(b) we have that $\mathbf{p} \cap J\Delta$ is an H-prime of $J\Delta$ and $\mathbf{q} \leq \mathbf{p}$ is prime in JH.

Set $Z = \zeta_1(H)$. Since Z is central, $\mathbf{p} \cap JZ$ is a prime ideal of JZ and trivially

$$Z \cap (1 + (\mathbf{p} \cap JZ)) \leq H \cap (1 + \mathbf{p})$$

and $(G : N_G(\mathbf{p} \cap JZ))$ are finite. Thus 7.10 yields that

$$\mathbf{p} \cap JZ = (\mathbf{p} \cap J(\Delta \cap Z))JZ \leq \mathbf{q}.$$

Consequently $\mathbf{p} \cap JZ = \mathbf{q} \cap JZ$.

Let $x \mapsto x^\wedge$ denote the canonical map of JH onto $JH/\mathbf{q}$. Suppose $\alpha \in JZ \backslash \mathbf{q}$ and $\beta \in JH$ satisfy $\alpha\beta \in \mathbf{q}$. Then $\alpha JH\beta \subseteq \mathbf{q}$ and the primeness of $\mathbf{q}$ yields that $\beta \in \mathbf{q}$. Thus $(JH)^\wedge$ is torsion-free as $(JZ)^\wedge$-module. In particular $(JZ)^\wedge$ is an integral domain.

Denote the quotient field of $(JZ)^\wedge$ by K. Then $(JH)^\wedge$ embeds into $Q = K \otimes_{(JZ)^\wedge} (JH)^\wedge$ and after the obvious identification $Q = K.(JH)^\wedge$. If $\mathbf{a}$ is an ideal of Q then $\mathbf{a} = K(\mathbf{a} \cap (JH)^\wedge)$. In particular Q is a prime ring since $(JH)^\wedge$ is a prime ring ($\mathbf{q}$ is a prime ideal). Also since $(H : Z)$ is finite, so Q is a finite-dimensional K-algebra. By Artin-Wedderburn theory Q is a simple ring. Thus $K.\mathbf{p}^\wedge$ is $\{0\}$ or Q. But $\mathbf{p} \cap JZ = \mathbf{q} \cap JZ$, so $\mathbf{p}^\wedge \cap (JZ)^\wedge = \{0\}$ and $K.\mathbf{p}^\wedge \cap (JZ)^\wedge = \{0\}$. Therefore $K.\mathbf{p}^\wedge = \{0\}$, $\mathbf{p}^\wedge = \{0\}$ and $\mathbf{p} = \mathbf{q}$ as required. $\qquad\square$

8.3 Lemma Let G be a polycyclic group, T its maximum finite normal subgroup, H/T the Fitting subgroup of G/T and A/T the centre of H/T. Then $A = \Delta_G(H)$.

The A here is a special case of what is often called the Zaleskii subgroup of a soluble group, see Passman (1977) or Shirvani and Wehrfritz (1986).

Proof If $a \in A$ and $h \in H$, then $a^h \in aT$, which is a finite set. Thus $(H : C_H(a))$ is finite and $A \le \Delta_G(H) \le \Delta_G(H/T)$. Thus we may assume that T is trivial. But then H is torsion-free, so we can choose an embedding of G into some $GL(n, \mathbf{Z})$ with H unipotent (see the proof of 4.11, especially 4.8.1). Every unipotent linear group in characteristic zero is connected (see Wehrfritz 1973a, 6.4 and 6.6), so H has no proper centralizer of finite index and hence $\Delta_G(H) = C_G(H)$. But A is the centre of the Fitting subgroup H of the soluble group G and $C_G(H) \le H$ by 5.6. Therefore $A = C_G(H)$. $\qquad\square$

Recall that a group G is said to be orbitally sound if H^G/H_G is finite for all subgroups H of G with the index $(G : N_G(H))$ finite; here $H^G = \langle H^g : g \in G \rangle$ and $H_G = \bigcap_{g \in G} H^g$.

8.4 Lemma (Roseblade 1978) Let G be an orbitally sound group, let K be a group of operators on G and suppose $\Delta = \Delta_G(K)$ is polycyclic-by-finite and normal in G. Let J be a commutative Noetherian ring and $\mathbf{q}$ a K-invariant G-prime of $J\Delta$ with $\Delta \cap (1 + \mathbf{q})$ finite. Let $x \mapsto x^\wedge$ denote the natural map of JG onto $JG/\mathbf{q}JG$. Then $\Delta_{(JG)^\wedge}(K) = (J\Delta)^\wedge$.

Proof Let T be a transversal of Δ to G. Then $JG = \bigoplus_{t \in T} J\Delta t$ and $(JG)^\wedge = \bigoplus_T (J\Delta t)^\wedge$. Let $\alpha \in \Delta_{(JG)^\wedge}(K) \backslash \{0\}$ and $\gamma \in C_K(\alpha)$. Now $\alpha = \alpha_1 + \alpha_2 + \cdots + \alpha_r$ say where α_i lies in $(J\Delta t_i)^\wedge \backslash \{0\}$ and the t_i are distinct members of T. Then $(\Delta t_i)^\gamma = \Delta t$ for some t in T and $\alpha_i^\gamma \in (J\Delta t)^\wedge$. Hence $\alpha_i^\gamma = \alpha_{i'}$ for some i'. Thus each α_i lies in $\Delta_{(JG)^\wedge}(K)$ and we may assume that $r = 1$. Thus let $\alpha^\wedge t^\wedge \in \Delta_{(JG)^\wedge}(K)$, where $\alpha \in J\Delta$ and $t \in T$.

There exists a subgroup K_0 of K of finite index centralizing $\alpha^\wedge t^\wedge$ and Δ. Then K_0 stabilizes the series $\langle 1 \rangle \le \Delta \le \Delta \langle t \rangle$. By 1.20 we have $[\Delta \langle t \rangle, K_0] \le \zeta_1(\Delta)$. Let $\gamma \in K_0$. Then $\alpha^\gamma = \alpha$ and $(\alpha^\wedge t^\wedge)^\gamma = \alpha^\wedge t^\wedge$, so $\alpha(t^\gamma - t) \in \mathbf{q} JG$ and

$$\alpha(t^\gamma t^{-1} - 1) \in \mathbf{q} JG \cap J\Delta = \mathbf{q}.$$

Note that $J\Delta$ is Noetherian. By 8.1(c) we have $\mathbf{q} = \mathbf{q}_1 \cap \mathbf{q}_2 \cap \cdots \cap \mathbf{q}_r$, where the $\mathbf{q}_i$ form a G-orbit of primes of $J\Delta$. Since $\alpha^\wedge \ne 0$, we may assume that α does not lie in $\mathbf{q}_1$. If $\gamma \in K_0$, then $\alpha(t^\gamma t^{-1} - 1) \in \mathbf{q}_1$ and $t^\gamma t^{-1}$ is central in Δ, so primeness yields that $t^\gamma t^{-1} - 1 \in \mathbf{q}_1$ for all γ in K_0. Therefore K_0, which centralizes Δ and $\mathbf{q}_1$, normalizes $Q_1 t$ for $Q_1 = \Delta \cap (1 + \mathbf{q}_1)$. Now

$$(Q_1)_G = \bigcap_{g \in G} Q_1^g = \Delta \cap \left(1 + \bigcap_{g \in G} \mathbf{q}_1^g \right) = \Delta \cap (1 + \mathbf{q}),$$

which is finite by hypothesis. Since G is orbitally sound Q_1 is finite (note that $N_G(Q_1)$ has finite index in G since Δ is normal in G). Thus $\{t^\gamma : \gamma \in K_0\} \subseteq Q_1 t$ is finite and $t \in \Delta$, as required. $\qquad\square$

8.5   Theorem (Roseblade 1978) Let J be a commutative ring and G an orbitally sound polycyclic group. If $\mathbf{p}$ is a prime ideal of JG, set $P = G \cap (1 + \mathbf{p})$ and $D/P = \Delta(G/P)$. Then $\mathbf{p} = (\mathbf{p} \cap JD)JG$.

Of course we use the notation $\Delta_G(G/P)$ for D as in 8.5.

Proof Assume first that $J = F$ a field. Since images of orbitally sound groups are orbitally sound and $P - 1 \subseteq \mathbf{p} \cap FD$, we may assume that $P = \langle 1 \rangle$. Let T be the maximum finite normal subgroup of G, H/T the Fitting subgroup of G/T and A/T the centre of H/T. By 8.1(a) and (c) there exists a finite G-orbit $\mathbf{q}_1, \mathbf{q}_2, \dots, \mathbf{q}_r$ of prime ideals of FA such that

$$\mathbf{p} \cap FA = \mathbf{q}_1 \cap \mathbf{q}_2 \cap \cdots \cap \mathbf{q}_r.$$

Clearly G also permutes transitively the $Q_i = A \cap (1 + \mathbf{q}_i)$, so that for each fixed j we have

$$\bigcap_{g \in G} Q_j^g = \bigcap_i Q_i = A \cap (1 + \mathbf{p}) \subseteq G \cap (1 + \mathbf{p}) = \langle 1 \rangle.$$

Since G is orbitally sound each Q_j is finite. Trivially each $(G : N_G(\mathbf{q}_j))$ is finite, so 8.2 yields for each j that

$$\mathbf{q}_j = (\mathbf{q}_j \cap F\Delta_A(G))FA = (\mathbf{q}_j \cap FD)FA,$$

the latter by 8.3 since $\Delta_A(G) = \Delta_G(H) \cap \Delta_G(G) = \Delta(G) = D$. Thus

$$\mathbf{p} \cap FA = \bigcap_i (\mathbf{q}_i \cap FD)FA = \left(\bigcap_i \mathbf{q}_i \cap FD \right) FA = (\mathbf{p} \cap FD)FA.$$

Set $\mathbf{q} = (\mathbf{p} \cap FD)FG$ and assume that $\mathbf{q} < \mathbf{p}$. By 6.6 there exists α in $\mathbf{p} \backslash \mathbf{q}$ such that $\alpha^g - \alpha \in \mathbf{q}$ for every g in the Fitting subgroup K of G. Now $H \geq K \geq C_H(T)$, so $(H : K)$ is finite and $\Delta_G(K) = \Delta_G(H) = A$ by 8.3. Also $\mathbf{p} \cap FA$ is a K-invariant G-prime of FA with $A \cap (1 + (\mathbf{p} \cap FA)) = \langle 1 \rangle$. Thus 8.4 yields that $\alpha \in FA + \mathbf{q}$; hence

$$\alpha \in \mathbf{p} \cap (FA + \mathbf{q}) = (\mathbf{p} \cap FA) + \mathbf{q} = \mathbf{q}.$$

This contradiction of the choice of α completes the proof the $\mathbf{p} = \mathbf{q}$ in this special case.

Now consider the general case. Again we may assume that $P = \langle 1 \rangle$ and also that $\mathbf{p} \cap J = \{0\}$. Then J is an integral domain; let F be its quotient field. If $\mathbf{a}$ is any ideal of FG then $\mathbf{a} = F(\mathbf{a} \cap JG)$ and if $\alpha \in F\mathbf{p} \cap JG$, say $\alpha = \beta \gamma^{-1}$ for $\beta \in \mathbf{p}$ and $\gamma \in J$, then $\alpha\gamma \in \mathbf{p}$ and γ is central and not in $\mathbf{p}$, so $\alpha \in \mathbf{p}$. Thus $F\mathbf{p} \cap JG = \mathbf{p}$. Consequently $F\mathbf{p}$ is a prime ideal of FG and $G \cap (1 + F\mathbf{p}) = \langle 1 \rangle$. By the above we have

$$F\mathbf{p} = (F\mathbf{p} \cap FD)FG.$$

Let $\alpha \in JG \cap (F\mathbf{p} \cap FD)FG$, say $\alpha = \sum_{t \in T} \alpha_t t$, where the $\alpha_t \in JD$ and T is a transversal of D to G. There exists $\eta \neq 0$ in J with $\alpha\eta \in (\mathbf{p} \cap JD)JG$. Thus each $\alpha_t\eta$ lies in $\mathbf{p}$, a prime and η is central and not in $\mathbf{p}$. Hence each α_t lies in $\mathbf{p} \cap JD$ and $\alpha \in (\mathbf{p} \cap JD)JG$. It now follows that $\mathbf{p} = JG \cap F\mathbf{p} = (\mathbf{p} \cap JD)JG$.   □

We have yet to prove 5.16 that every polycyclic contains an orbitally sound normal subgroup of finite index. This is a special case of 8.6 below by 4.8, the Auslander Embedding Theorem.

8.6   Theorem If G is a soluble connected subgroup of $GL(n, \mathbf{Z})$, then G is orbitally sound.

It follows from 8.6 that a finitely generated nilpotent group G is orbitally sound; for if T is the torsion subgroup of G, then G/T is isomorphic to a unipotent linear group over $\mathbf{Z}$, which is always connected and hence orbitally sound. Thus T is finite and G/T is orbitally sound, so G is orbitally sound. This result also follows from isolator theory; for if H is a subgroup of G with $(G : N_G(H))$ finite, then $N_G i_G(H) = i_G N_G(H) = G$, see 5.13. Thus $H \leq H^G \leq i_G(H)$, so $(H^G : H)$ is finite. But $(H^G : H^g) = (H^G : H)$ for any g in G and H^G being finitely generated has only a finite number of subgroups of this index. Consequently $(H^G : H_G)$ is finite and G is orbitally sound. To prove 8.6 we need a few lemmas. In these lemmas words such as closed and closure refer to the Zariski topology, see Chap. 5.

8.6.1   Let F be a field, G a subgroup of $Tr(n, F)$ and H a subgroup of G such that $U = u(H)$ (the unipotent radical of H) is closed in G. Then $N_G(H)$ is closed in G.

Proof For $[N_G(H), H] \leq G' \cap H \leq U$. Since U is closed in G, 4.16 yields that $[N_G(H)^\wedge, H] \leq U$, closures being taken in G. Thus $N_G(H)^\wedge \leq N_G(H)$ and hence $N_G(H)$ is closed in G.   □

8.6.2   Let V be the closure in $GL(n, \mathbf{Z})$ of the unipotent subgroup U of $GL(n, \mathbf{Z})$. Then U has finite index in V.

Proof Initially we work over the rationals $\mathbf{Q}$. We may assume that $U \le Tr_1(n, \mathbf{Q})$. Now the closure of U in $GL(n, \mathbf{Q})$ is just the isolator of U in $Tr_1(n, \mathbf{Q})$: for a formal proof of this see 3.2 of Wehrfritz (1974) or Proposition 8, p. 167 of Segal (1983), but basically the reason is that U has a central series of finite length whose factors are non-trivial subgroups of the additive group of $\mathbf{Q}$ and if X is one such, then $\mathbf{Q}/X$ is periodic, so $i_\mathbf{Q}(X) = \mathbf{Q}$, and $\mathbf{Q}$ is the closure of X, the proper closed subsets of $\mathbf{Q}$ being finite.

Thus $V = i_V(U)$. Moreover V is and all its subgroups are finitely generated nilpotent groups by 4.4. Consequently U is subnormal in V and $(V : U)$ is finite. $\square$

8.6.3   Let F be a field, G a connected subgroup of $Tr(n, F)$, H a subgroup of G with $(G : N_G(H))$ finite and V the closure of $u(H)$ in G. Then HV is a normal subgroup of G.

Proof For $V \le HV \cap u(G) = u(HV) = (H \cap u(G))V = V$. Hence $N_G(HV)$ is closed in G by 8.6.1. Since G is connected and $N_G(H) \le N_G(HV)$ we have $G = N_G(HV)$. $\square$

Proof of 8.6 Note first that G is polycyclic by 4.4 and triangularizable by 4.13. Let H be a subgroup of G with $(G : N_G(H))$ finite. Let V denote the closure of $u(H)$ in G. Then $(V : u(H))$ is finite by 8.6.2, so $(HV : H)$ is finite. By 8.6.3 the subgroup HV is normal in G, so $H^G \le HV$ and $(H^G : H)$ is finite. Thus $(H^G : H_G)$ is finite and G is orbitally sound. $\square$

Let $\mathbf{Zr}$ denote the class of all polycyclic groups G such that for all commutative rings J and all prime ideals $\mathbf{p}$ in JG, we have $\mathbf{p} = (\mathbf{p} \cap JD)JG$, where D is as in 8.5. Thus by 8.5 all orbitally sound, polycyclic groups lie in $\mathbf{Zr}$. Not every polycyclic group is in $\mathbf{Zr}$. We refer the reader to Roseblade (1978) for a proof of the following.

8.7   Theorem (Roseblade 1978) If G is a polycyclic group, then $G \in \mathbf{Zr}$ if and only if whenever H is a subgroup of G with $(G : N_G(H))$ finite, then

$$\delta_G(H) = \{x \in G : [x, G_x] \subseteq H \text{ for some subgroup } G_x \text{ of } G \text{ of finite index}\}$$

is normal in G.

In 8.7 the set $\delta_G(H)$ is always a subgroup of G with its normalizer of finite index in G. Also H is clearly a subgroup of $\delta_G(H)$. Moreover G is orbitally sound if and only if every subgroup H of G, with $(G : N_G(H))$ finite and no other such subgroup of G containing H as a subgroup of finite index, is normal in G. (Some authors call a subgroup H of a polycyclic-by-finite group *orbital* if $(G : N_G(H))$ is finite and an *isolated orbital* if also no other orbital contains H as a subgroup of finite index. We are avoiding this use of the word isolated since it clashes with our earlier use. In the infinite dihedral group $\langle 1 \rangle$ is an isolated orbital subgroup in the above sense but is not an isolated subgroup.) However if G is just in **Zr**, then every such H is at least subnormal in G of depth at most 2.

8.8

(a) The wreath product $G = A wr C$ of the infinite cyclic group A by the cyclic group C of order 2 is polycyclic and lies in **Zr**, but is not orbitally sound.

(b) The permutational wreath product $A wr_3 S$, where A is infinite cyclic and $S =$ Sym(3), is polycyclic but is not in **Zr**. In particular G is also not orbitally sound.

Proof (a) Let $G = (\langle a \rangle \times \langle b \rangle) \langle c \rangle$, where a and b have infinite order, c has order 2 and c interchanges a and b. Clearly G is polycyclic. Set $H = \langle a \rangle$. Then $N_G(\langle a \rangle) = \langle a, b \rangle$, which is normal and has finite index 2 in G. Thus $H^G = \langle a, b \rangle$. But $H_G = \langle 1 \rangle$ and H^G / H_G is infinite. Therefore G is not orbitally sound. Trivially $[\langle a, b \rangle, \langle a, b \rangle] = \langle 1 \rangle \leq H$ for any subgroup H of G, so in the notation of 8.7 above $\delta_G(H) \geq \langle a, b \rangle$ and hence is normal in G. Consequently $G \in$ **Zr**.

(b) Let F be any field with an element λ of infinite order. Then

$$\left\langle \begin{pmatrix} \lambda & 0 & 0 \\ 0 & 1 & 0 \\ 0 & 0 & 1 \end{pmatrix}, \begin{pmatrix} 0 & 1 & 0 \\ 0 & 0 & 1 \\ 1 & 0 & 0 \end{pmatrix}, \begin{pmatrix} 0 & 1 & 0 \\ 1 & 0 & 0 \\ 0 & 0 & 1 \end{pmatrix} \right\rangle$$

is isomorphic to G and we take it to be G. Clearly $F[G]$ is the full matrix algebra $F^{3 \times 3}$, which is simple, and there is a homomorphism of the group algebra FG onto $F[G]$ that is the identity on G. Thus FG has a prime ideal **p** with $FG/\mathbf{p} \cong F^{3 \times 3}$ and $G \cap (1 + \mathbf{p}) = \langle 1 \rangle$. Also $\Delta(G)$ is the base group of G, which is just the diagonal subgroup of G. This latter clearly spans the full diagonal algebra. If $\mathbf{p} = (\mathbf{p} \cap F\Delta(G))FG$, then Sym(3) is linearly independent over the diagonal algebra, which has dimension 3 over F. Thus $6 = 3^2/3$, a contradiction that confirms that G does not lie in **Zr** and hence that G is not orbitally sound. ☐

We now consider the prime length of the group algebra of a polycyclic group, the *prime length* of a ring R being the maximum length of a finite chain of prime ideals of the ring (the ring itself not being counted as a prime ideal), assuming this exists. Let F be any field and G an orbitally sound polycyclic group. We claim that the prime length of FG is at most one more than the square of the Hirsch number $h(G)$ of G (we will see that it is at most $h(G)$ a little later). In particular the prime length of FG exists.

First consider a ring $R = \bigoplus_{t \in T} Jt$ that is a free J-module of finite rank over its central integral domain J. If K denotes the quotient field of J, then $K \otimes_J R = \bigoplus_T Kt = KR$ is a finite-dimensional K-algebra and hence has prime length 0. If **a** is an ideal of KR, then $\mathbf{a} = K(\mathbf{a} \cap R)$. If **p** is a prime ideal of R with $J \cap \mathbf{p} = \{0\}$, then $K\mathbf{p} < KR$. Also if **a** and **b** are ideals of KR with $\mathbf{ab} \leq K\mathbf{p}$, then $(\mathbf{a} \cap R)(\mathbf{b} \cap$

$R) \le K\mathbf{p} \cap R = \mathbf{p}$. If $\mathbf{a} \cap R \le \mathbf{p}$, then $\mathbf{a} \le K\mathbf{p}$, and similarly with $\mathbf{b}$. Hence $K\mathbf{p}$ is a prime ideal of KR. Thus if $\mathbf{q} \ge \mathbf{p}$ is a prime of R with $J \cap \mathbf{q} = \{0\}$, then $K\mathbf{q} = K\mathbf{p}$ and hence $\mathbf{q} = \mathbf{p}$.

Now consider prime ideals $\mathbf{p} < \mathbf{q}$ of the group algebra FG. We may factor out by $P = G \cap (1 + \mathbf{p})$ and assume that $P = \langle 1 \rangle$. Set $Q = G \cap (1 + \mathbf{q})$. If Q is infinite, by induction on the Hirsch length of G/Q we have that $FG/(Q-1)FG$ has prime length at most $1 + h(G/Q)^2 \le h(G)^2$. Thus any chain of prime ideals of FG that starts $\mathbf{p} < \mathbf{q}$ has length at most $2 + h(G/Q)^2 \le 1 + h(G)^2$.

Now assume Q is finite. Let $\Delta = \Delta(G)$ and note that $Q \le \Delta$ and $\Delta/Q = \Delta(G/Q)$. By 8.5 we have that $\mathbf{p} = (\mathbf{p} \cap F\Delta)FG$ and $\mathbf{q} = (\mathbf{q} \cap F\Delta)FG$. Also by 8.1 we have

$$\mathbf{p} \cap F\Delta = \bigcap_i \mathbf{p}_i \quad \text{and} \quad \mathbf{q} \cap F\Delta = \bigcap_j \mathbf{q}_j,$$

where the $\mathbf{p}_i$ are the minimal primes of $F\Delta$ over $\mathbf{p} \cap F\Delta$, permuted transitively by G, and similarly for the $\mathbf{q}_j$ and $\mathbf{q} \cap F\Delta$. Since $\mathbf{p} < \mathbf{q}$ for each j there exists some $i(j)$ with $\mathbf{p}_{i(j)} < \mathbf{q}_j$. Consequently any chain of prime ideals $\mathbf{r}$ of FG starting with $\mathbf{p}$ and with each $G \cap (1 + \mathbf{r})$ finite has length at most the prime length of $F\Delta$. There exists a central free abelian subgroup Z of Δ with Δ/Z finite. Let $\mathbf{r}$ be any prime ideal of $F\Delta$. Now $\mathbf{r} \cap FZ$ is a prime ideal of FZ and $FZ/(\mathbf{r} \cap FZ)$ is an integral domain. Thus $F\Delta/(\mathbf{r} \cap FZ)F\Delta$ satisfies the hypotheses of our remark at the start of this proof. Consequently the prime length of $F\Delta$ is equal to the prime length of FZ. If follows from Kaplansky (1970), top of p. 109, that the latter is equal to $h(Z)$. Hence any chain of prime ideals of FG starting with $\mathbf{p}$ has length at most 0 if $h(G) = 0$, at most 2 if $h(G) = 1$ and at most

$$h(Z) + 2 + (h(G) - 1)^2 \le 1 + h(G)^2,$$

for $h(G) \ge 2$. The claim follows.

With more care one can reduce $1 + h(G)^2$ in the above to $h(G)$. An alternative approach to this, indeed to a generalization, uses the notion of the Krull dimension of a ring, see McConnell and Robson (1987). Let G be a polycyclic-by-finite group and J any right Noetherian ring with finite (right) Krull dimension k. For example J could be any field, when $k = 0$, or the integers $\mathbf{Z}$, when $k = 1$. By Proposition 6.6.1 of McConnell and Robson (1987) the group ring JG has Krull dimension $h(G) + k$. Also by Lemma 6.4.5 of the same work any chain of prime ideals of the right Noetherian ring JG has length at most its Krull dimension. Thus we have the following.

8.9   Theorem Let G be a polycyclic-by-finite group and F any field. Then any chain of prime ideals in FG has length at most $h(G)$ and any chain of prime ideals in $\mathbf{Z}G$ has length at most $h(G) + 1$.

Since $G \cap (1 + \mathbf{a})$ is a normal subgroup of G for any ideal $\mathbf{a}$ of FG, one would actually expect to do better than $h(G)$. This is indeed the case, but we need first to define a new invariant for our group G.

If G is a polycyclic group, the *plinth length* $pl(G)$ is the maximum integer r such that G has a subgroup H of finite index in G with a normal series

$$\langle 1 \rangle = H_0 < H_1 < \cdots < H_i < \cdots < H_r = H$$

with each H_i/H_{i-1} free abelian and rationally irreducible as H-module. Obviously $pl(G)$ is bounded by the Hirsch number $h(G)$ and in particular exists. The following generalizes our remarks above; see Roseblade (1978) for the proof.

8.10   Theorem (Roseblade 1978, Theorem H1) Let F be a field and G a polycyclic group. Then the maximum length of a chain of prime ideals of the group algebra FG is the plinth length $pl(G)$ of G.

The following two results are Corollary H3 and Theorem H4 of the same paper and again see Roseblade (1978) for the proofs. Here $ht(\mathbf{p})$ of a prime ideal $\mathbf{p}$ denotes the height of $\mathbf{p}$; that is, the maximum length of a chain of prime ideals

$$\mathbf{p} = \mathbf{p}_0 > \mathbf{p}_1 > \cdots > \mathbf{p}_r$$

descending from $\mathbf{p}$.

8.11   Theorem (Roseblade 1978) Let F be a field and suppose $G \in \mathbf{Zr}$. If $\mathbf{p} < \mathbf{q}$ are prime ideals of FG with no prime ideal strictly between them, then

$$ht(\mathbf{q}) = 1 + ht(\mathbf{p}).$$

8.12   Theorem (Roseblade 1978) Let F be a locally finite field and suppose $G \in \mathbf{Zr}$. If $\mathbf{p}$ is any prime ideal of FG, then $FG/\mathbf{p}$ has prime length $pl(G) - ht(\mathbf{p})$.

Incidentally if we drop the condition in 8.12 that F is locally finite, then 8.12 becomes false, even for finitely generated nilpotent groups. Now 8.10 has generalized and strengthened our remarks above from orbitally sound polycyclic groups to just polycyclic groups. We have already seen that the orbitally sound condition in our main theorem (8.5) cannot be dropped, see 8.8. However not all is lost. The following is essentially a combination of Theorems I, II and III of Lorenz and Passman (1981).

8.13   Theorem (Lorenz and Passman 1981) Let J be a commutative ring and G a polycyclic-by-finite group. If $\mathbf{p}$ is a prime ideal of JG there exists a subgroup H of G, with $(G : N_G(H))$ finite and no other such subgroup of G containing H as a subgroup of finite index, such that $\mathbf{p} = \mathbf{q}_G = \bigcap_{g \in G} \mathbf{q}^g JG$ for some prime ideal $\mathbf{q}$ of $J\delta_G(H)$ with $\delta_G(H) \cap (1 + \mathbf{q})$ of finite index in H. Moreover H is unique up to conjugation, given H the ideal $\mathbf{q}$ is unique up to conjugation by $\delta_G(H)$ and given any such H and $\mathbf{q}$ the ideal $\mathbf{q}_G$ is always a prime ideal of JG.

For further developments of this work of Lorenz and Passman (1981), see Letzter and Lorenz (1999).

The following indicates a second possible direction to travel from 8.5. It is derived from work of A.E. Zalesskii, but for details see Passman (1977) 11.4.10 or Shirvani and Wehrfritz (1986) 5.3.7(c).

8.14   Let J be a ring and G a polycyclic group and let $A/\tau(G)$ denote the centre of the Fitting subgroup of $G/\tau(G)$. Suppose $\mathbf{a}$ is an ideal of JG such that for every infinite subgroup B of A the left annihilator of the image of $B - 1$ in $JG/\mathbf{a}$ is $\{0\}$ (that is, if $\alpha \in JG$ with $\alpha(B - 1) \subseteq \mathbf{a}$, then $\alpha \in \mathbf{a}$). Then $\mathbf{a} = (\mathbf{a} \cap JA)JG$.

In 8.14 if G is a finitely generated nilpotent group and $\mathbf{p}$ is a prime ideal of JG, then it is easy to see for P and D as in 8.5 that, modulo $(P - 1)JG$, the group G and the ideal $\mathbf{p}$ satisfy 8.14 and so $\mathbf{p} = (\mathbf{p} \cap JD)JG$. Here we have not needed to assume that J is commutative as we would if we obtained this from 8.5 (recall G is orbitally sound by the remarks after the statement of 8.6). See Shirvani and Wehrfritz (1986) 5.3.13.

Chapter 9
The Structure of Modules over Polycyclic Groups

In many ways this chapter is the culmination of much of the work we have done in Chaps. 6, 7 and 8. We are especially interested here in the structure of a finitely generated module over a polycyclic group. We then use this information to prove that a finitely generated abelian-by-polycyclic-by-finite group is residually finite.

Hall (1959) had earlier proved that finitely generated abelian-by-nilpotent-by-finite groups are residually finite. This work of Hall required a careful study of finitely generated modules over finitely generated nilpotent groups. Much of this chapter is devoted to Roseblade's extension of this work of Hall. This is not a simple extension; indeed the polycyclic case is far more involved and requires a far wider range of techniques.

We then conclude the chapter with a summary of further results in this area.

9.1 Theorem Let $R = S[G]$ be a ring generated as a ring by its right Noetherian subring S and the polycyclic-by-finite subgroup G of its group of units normalizing S. Let M be a finitely generated R-module and N any finitely generated S-submodule of M. Then there exists an ascending series

$$N = M_0 \leq M_1 \leq \cdots \leq M_\lambda \leq \cdots \leq M_\mu = M$$

of S-submodules of M such that the factors $M_{\lambda+1}/M_\lambda$ are cyclic S-modules that fall into a finite number of G-conjugacy classes.

The history of this theorem is, I think, as follows. Hall (1959) essentially has 9.1 in the case where S and G commute. Roseblade (1973a) has 9.1 with $N = \{0\}$. This is, in fact, the main case. The full result is in Segal (1977). The proof of 9.1 below is essentially Roseblade's.

Proof First consider the case where $G = \langle x \rangle$ is infinite cyclic. Then

$$R = \sum_{-\infty < n < \infty} Sx^n.$$

B.A.F. Wehrfritz, *Group and Ring Theoretic Properties of Polycyclic Groups*,
Algebra and Applications 10,
DOI 10.1007/978-1-84882-941-1_9, © Springer-Verlag London Limited 2009

Let U be any finitely generated S-submodule of M containing N with $UR = M$. For each $n = 0, 1, 2, \ldots$ set $U_n = \sum_{0 \leq i \leq n} Ux^i$ and $V = \bigcup_{n \geq 0} U_n$. Clearly

$$U = U_0 \leq U_1 \leq U_2 \leq \cdots$$

and since x normalizes S, each U_n is an S-submodule of M. For $n = 0, 1, 2, \ldots$ set

$$V_n = \sum_{-n \leq i < \infty} Ux^i.$$

Then V_n is also an S-submodule of M with $V = V_0 \leq V_1 \leq V_2 \leq \cdots$ and $M = \bigcup_{n \geq 0} V_n$.

Clearly $V_n = V_{n-1} + Ux^{-n}$ for $n \geq 1$, so

$$V_n/V_{n-1} \cong_S Ux^{-n}/(V_{n-1} \cap Ux^{-n}) = Ux^{-n}/(Vx^{-(n-1)} \cap Ux^{-n}),$$

and the latter is a G-conjugate of $U/(Vx \cap U)$. In particular it is finitely generated as S-module. Also $U_n = U_{n-1} + Ux^n$ for $n \geq 1$, so

$$U_n/U_{n-1} \cong_S Ux^n/(U_{n-1} \cap Ux^n),$$

which is a G-conjugate of $U/(U_{n-1}x^{-n} \cap U)$. Now $U_{n-1}x = \sum_{1 \leq i \leq n} Ux^i \leq U_n$, so $U_{n-1}x^{-n} \leq U_n x^{-(n+1)}$. Since U is Noetherian as S-module, there exists m such that $U_{n-1}x^{-n} \cap U = U_{m-1}x^{-m} \cap U$ for all $n \geq m$. Hence if

$$\mathbf{X} = \{U/(Vx \cap U), U/(U_i x^{-i-1} \cap U) : i = 0, 1, 2, \ldots, m\},$$

then $M/U \in P'\mathbf{X}^G$, meaning that M/U has an ascending series of S-modules running from $\{0\}$ to M/U whose factors are G-conjugates of elements of $\mathbf{X}$. (Here P' denotes the ascendant operator, a generalization of P, the poly operator introduced in Chap. 2.) Since each element of $\mathbf{X}$ is S-Noetherian, there exists a finite set $\mathbf{Y}$ of cyclic S-modules such that $\mathbf{X} \cup \{U/N\} \subseteq P\mathbf{Y}$. Then $M/N \in P'\mathbf{Y}^G$ and the theorem is proved in this case.

In general G has a series whose factors are finite or infinite cyclic. Hence we may assume that there is a normal subgroup G_1 of G for which the theorem holds, such that G/G_1 is finite or infinite cyclic. Set $S_1 = S[G_1]$ and suppose G/G_1 is infinite cyclic. By the above there exists a finite set $\mathbf{X}_1$ of cyclic S_1-modules such that $M/NG_1 \in P'\mathbf{X}_1^G$. By hypothesis there exists a finite set $\mathbf{X}$ of cyclic S-modules such that $\{NG_1/N\} \cup \mathbf{X}_1 \subseteq P'\mathbf{X}^{G_1}$. Then $M/N \in P'(P'\mathbf{X}^{G_1})^G \subseteq P'\mathbf{X}^G$ and the theorem is proved in this case.

Finally we must consider the case where G/G_1 is finite. In this case M is finitely generated as S_1-module, so for some finite set $\mathbf{X}$ of cyclic S-modules we have $M/N \in P'\mathbf{X}^{G_1} \subseteq P'\mathbf{X}^G$. The proof is now complete. $\qquad\square$

9.2   Corollary (Roseblade 1973a) Let J be a commutative Noetherian domain, let G be a polycyclic-by-finite group and set $R = JG$. Suppose A is a free abelian normal subgroup of G and set $S = JA$, regarded as a subring of R. If M is a finitely R-generated R-module, then there exists a free S-submodule F of M and a non-zero ideal $\mathbf{a}$ of S such that every finitely generated S-submodule of M/F is annihilated by a product of a finite number of G-conjugates of $\mathbf{a}$.

For example, let $J = \mathbf{Z}$, let $A = \langle 1 \rangle$ and let M be a $\mathbf{Z}$-torsion-free finitely generated G-module. Then (Hall 1959) there is a free abelian subgroup F of M and a finite set π of primes such that M/F is additively a π-group (for π take the set of prime divisors of the order of $\mathbf{Z}/\mathbf{a}$). There is a similar corollary (again due to Hall 1959) with $J = \mathbf{F}_p$, the field of p-elements for some prime p, and A an infinite cyclic, central subgroup of G, so again $S = \mathbf{F}_p A$ is a very nice principal ideal domain.

Proof By 9.1 there is an ascending series $\{M_\alpha\}$ of S-submodules of M with $M_0 = \{0\}$ such that for some finite set $\mathbf{X}$ of cyclic S-modules each $M_{\alpha+1}/M_\alpha \in \mathbf{X}^G$. Let $\mathbf{a}$ denote the product of those annihilators $\mathrm{Ann}_S X$ that are non-zero as X runs over $\mathbf{X}$, with $\mathbf{a} = S$ if no such annihilator exists. Now S is a commutative domain, so $\mathbf{a}$ is a non-zero ideal of S. For each α choose x_α in M with $M_{\alpha+1} = M_\alpha + x_\alpha S$. Let

$$F = \sum (x_\alpha S : M_{\alpha+1}/M_\alpha \text{ is isomorphic to } S).$$

Then F has an ascending series of S-submodules with all factors isomorphic to S. Consequently F is a free S-submodule of M.

Let U be a finitely S-generated submodule of M and let $\lambda(1)$ be minimal subject to $U \leq M_{\lambda(1)} + F$. Now $\lambda(1)$ is not a limit ordinal since U is finitely generated, so set $\lambda(1) = \mu(1) + 1$. If $X_1 = M_{\lambda(1)}/M_{\mu(1)}$ is isomorphic to S, then $M_{\lambda(1)} + F = M_{\mu(1)} + F$ by the definition of F. Hence $\mathbf{b} = \mathrm{Ann}_S X_1$ is non-zero. There exists $g(1)$ in G with $M_{\lambda(1)} \mathbf{a}^{g(1)} \leq M_{\lambda(1)} \mathbf{b} \leq M_{\mu(1)}$. Thus $U \mathbf{a}^{g(1)} \leq M_{\mu(1)} + F$. Also $U \mathbf{a}^{g(1)}$ is finitely S-generated, since S is Noetherian. Consequently in the same way there exists a $g(2)$ in G with $U \mathbf{a}^{g(1)} \mathbf{a}^{g(2)}$ contained in $M_{\mu(2)} + F$ for some $\mu(2) < \mu(1)$. This process halts after a finite number of steps. $\qquad\square$

9.3 Corollary Assume the notation and hypotheses of 9.2. Let $\mathbf{m}$ be a maximal ideal of S such that $\mathbf{m}$ does not contain $\mathbf{a}^g$ for any g in G. Then

(a) $M = F + M\mathbf{m}$ and $M\mathbf{m} \cap F = F\mathbf{m}$, and
(b) $\mathbf{m}_G = \bigcap_{g \in G} \mathbf{m}^g = \{0\}$ whenever M is S-torsion-free and R-irreducible.

Proof (a) Let U be a finitely generated S-submodule of M. By 9.2 there exist elements $x(1), x(2), \ldots, x(n)$ of G with $U\mathbf{b} \leq F$ for $\mathbf{b} = \mathbf{a}^{x(1)}.\mathbf{a}^{x(2)}.\ldots.\mathbf{a}^{x(n)}$. Since $\mathbf{a}^{x(i)}$ does not lie in $\mathbf{m}$ for each i, we have that $\mathbf{b}$ does not lie in $\mathbf{m}$. Hence $S = \mathbf{b} + \mathbf{m}$ and $U = US \leq F + U\mathbf{m}$. This is for all such U, so $M = F + M\mathbf{m}$. Also

$$U\mathbf{m} \cap F = (U\mathbf{m} \cap F)(\mathbf{b} + \mathbf{m}) \leq U\mathbf{b}\mathbf{m} + F\mathbf{m} = F\mathbf{m},$$

since S is commutative, so $M\mathbf{m} \cap F = F\mathbf{m}$.

(b) Since F is free and not zero (recall M is S-torsion-free and not zero), so $F > F\mathbf{m}$. Hence $M > M\mathbf{m}$ by part (a). Consequently $M\mathbf{m}_G = \{0\}$, since M here is R-irreducible. Again M is S-torsion-free, so $\mathbf{m}_G = \{0\}$.   □

9.4 Corollary (Segal 1977) Let J be a commutative Noetherian domain, G a polycyclic-by-finite group and M a finitely generated R-module for $R = JG$. Suppose H is a normal subgroup of G and N is a finitely generated S-submodule of M for $S = JH$. Then J has a non-zero ideal $\mathbf{a}$ such that for every maximal ideal $\mathbf{m}$ of J not containing $\mathbf{a}$ we have $M\mathbf{m} \cap N = N\mathbf{m}$.

Proof By 9.1 there is a finite set $\mathbf{X}$ of cyclic S-modules such that $M/N \in P'\mathbf{X}^G$. Apply 9.2 to each member of $\mathbf{X}$ with $S = J$. Thus there is a non-zero ideal $\mathbf{a}$ of J such that each member of $\mathbf{X}$ is J-free by $\mathbf{a}$-torsion. Then M/N contains a free J-submodule F/N such that M/F is an $\mathbf{a}$-torsion J-module.

Let $\mathbf{m}$ be a maximal ideal of J not containing $\mathbf{a}$. Then $J = \mathbf{m} + \mathbf{a}^i$ for each $i \geq 1$. If $x \in M$, there exists some i with $x\mathbf{a}^i \leq F$ and then $x\mathbf{m} \cap F = (x\mathbf{m} \cap F)(\mathbf{m} + \mathbf{a}^i) \leq F\mathbf{m}$. Therefore $F \cap M\mathbf{m} = F\mathbf{m}$. Now F/N is a free J-module, so $F = N \oplus E$ for some J-submodule E of F. Consequently

$$N \cap M\mathbf{m} = N \cap F\mathbf{m} = N \cap (N\mathbf{m} \oplus E\mathbf{m}) = N\mathbf{m},$$

as required.   □

Let R be any ring and M a right R-module. For X a subset of R and Y a subset of M, set

$$^*X = \{y \in M : yX = \{0\}\} \quad \text{and} \quad Y^* = \{x \in R : Yx = \{0\}\}.$$

$^*X$ is often called the victim of X in M. Y^* is just the annihilator $\mathrm{Ann}_R\, Y$ of Y in R, but this alternative notation is very convenient here. Note that if $RX \subseteq X$ then $^*X$ is a submodule of M and if $YR \subseteq Y$ then Y^* is an ideal of R. Let $\pi_R M$ denote the set of ideals of R that are maximal amongst all the ideals $\mathbf{a}$ of R with $^*\mathbf{a}$ non-zero. Frequently $\pi_R M$ is empty.

9.5 Lemma Assume the notation above. Then

(a) $\pi_R M$ consists only of prime ideals of R and
(b) $\sum_{\mathbf{p}} {}^*\mathbf{p} = \bigoplus_{\mathbf{p}} {}^*\mathbf{p}$, where $\mathbf{p}$ ranges over $\pi_R M$.

Proof (a) Suppose $\mathbf{a}$ and $\mathbf{b}$ are ideals of R with $\mathbf{ab} \leq \mathbf{p} \in \pi_R M$, where $\mathbf{p} \leq \mathbf{a} \cap \mathbf{b}$. If $^*\mathbf{a} \neq \{0\}$, then $\mathbf{p} \leq \mathbf{a}$ yields $\mathbf{a} = \mathbf{p}$ by the maximality of $\mathbf{p}$. If $^*\mathbf{a} = \{0\}$, then $\{0\} \neq (^*\mathbf{p})\mathbf{a} \leq {}^*\mathbf{b}$; whence $\mathbf{p} \leq \mathbf{b}$ yields $\mathbf{p} = \mathbf{b}$. Therefore $\mathbf{p}$ is prime.

(b) Let $\mathbf{p}_1, \mathbf{p}_2, \ldots, \mathbf{p}_r$ be distinct members of $\pi_R M$, where $1 < r < \infty$. It suffices to prove that the sum $^*\mathbf{p}_1 + \cdots + {}^*\mathbf{p}_r$ is direct. Since by (a) the ideal $\mathbf{p}_1$ is prime, $\mathbf{p}_1$ is strictly smaller than $\mathbf{q} = \mathbf{p}_1 + \mathbf{p}_2\mathbf{p}_3 \cdots \mathbf{p}_r$ and $^*\mathbf{q} = \{0\}$ by the maximality of $\mathbf{p}_1$. Consequently $(^*\mathbf{p}_1) \cap (^*\mathbf{p}_2 + \cdots + {}^*\mathbf{p}_r) = \{0\}$ and the lemma follows.   □

9.6 Let J be a ring, G a group, H a normal subgroup of G and set $R = JG$ and $S = JH$. Let M be any R-module. If $\mathbf{p} \in \pi_S M$ and if T is a right transversal of $N = N_G(\mathbf{p})$ to G, then

$$(^*\mathbf{p})R = \bigoplus_{t \in T} {}^*(\mathbf{p}^t) = \bigoplus_{t \in T} (^*\mathbf{p})t \cong (^*\mathbf{p}) \otimes_{JN} JG.$$

Proof If $x \in G$, then $(^*\mathbf{p})x.\mathbf{p}^x = \{0\}$ and hence $(^*\mathbf{p})x = {}^*(\mathbf{p}^x)$. Since $G = NT$ and since $^*\mathbf{p}$ is a J-submodule of M, we have

$$(^*\mathbf{p})R = \sum_{x \in G} (^*\mathbf{p})x = \sum_{x \in G} {}^*(\mathbf{p}^x) = \sum_{t \in T} {}^*(\mathbf{p}^t).$$

Now the $\mathbf{p}^t$ for t in T are distinct members of $\pi_S M$. Consequently the result follows from 9.5(b).   □

9.7 Corollary Assume the notation of 9.6. Then $U \mapsto UR = \bigoplus_{t \in T} Ut$ is a one-to-one inclusion preserving map from the set of JN-submodules of $^*\mathbf{p}$ into the set of R-submodules of M. In particular if M is R-Noetherian, then $^*\mathbf{p}$ is JN-Noetherian and if M is R-irreducible, then $^*\mathbf{p}$ is JN-irreducible.

All of 9.7 follows at once from 9.6. The point of all this is that if the index $(G : N)$ is infinite and if G is polycyclic-by-finite, then we can apply induction to the JN-module $^*\mathbf{p}$.

9.8 Theorem (Roseblade 1973a) Let F be a locally finite field and G a polycyclic-by-finite group. Then any irreducible FG-module M has finite dimension over F.

Proof For the given field F we induct on the Hirsch number of G. If G is finite every finitely generated FG-module has finite dimension over F, so assume G is infinite. Suppose G has a normal subgroup G_1 of finite index for which the theorem holds. Then M is also a finitely generated FG_1-module and so contains a maximal FG_1-submodule M_1. By hypothesis $\dim_F(M/M_1)$ is finite. Also $\{M_1 g : g \in G\}$ is finite and $\bigcap_{g \in G} M_1 g = \{0\}$. Thus the claim will follow in this case.

Replacing G by a suitable one of its normal subgroups of finite index we may now assume that G contains a non-trivial torsion-free abelian normal subgroup A upon which G acts $\mathbf{Q}$-irreducibly and connectedly (so A is a plinth of G). Set $R = FG$ and $S = FA \leq R$ and suppose that M is S-torsion-free. If $\mathbf{m}$ is a maximal ideal of S, then $\dim_F(S/\mathbf{m})$ is finite by Hilbert's Nullstellensatz and so $\dim_F(S/\mathbf{m}_G)$ is finite by 7.12. Hence $\mathbf{m}_G \neq \{0\}$. By 9.3(b) it follows that every maximal ideal of S contains some G-conjugate of the non-zero ideal $\mathbf{a}$ of 9.2 of S. This explicitly contradicts 7.17. Therefore M is not S-torsion-free.

In the notation of 9.5 and 9.6 we now have that $\pi_S M$ is not empty; choose $\mathbf{p}$ in $\pi_S M$. By 9.5(a) the ideal $\mathbf{p}$ is prime. Set $N = N_G(\mathbf{p})$. By 9.7 the FN-module $^*\mathbf{p}$ is irreducible. If N has infinite index in G then $\dim_F(^*\mathbf{p})$ is finite by induction. In this case $\dim_F(S/\mathbf{p})$ is also finite, since $S/\mathbf{p}$ acts faithfully on $^*\mathbf{p}$ by definition of $\pi_S M$. Now suppose N has finite index in G. Again we show that $\dim_F(S/\mathbf{p})$ is finite. Here A is a plinth for N. By 7.5 (Bergman's Theorem) either $\mathbf{p}$ is $\{0\}$ or $\dim_F(S/\mathbf{p})$ is finite. Clearly $\mathbf{p}$ cannot be $\{0\}$ and hence the claim is substantiated.

Since F is a locally finite field $A/(A \cap (1 + \mathbf{p}))$ is finite (for if $\dim_F(S/\mathbf{p}) = n$, then $A/(A \cap (1 + \mathbf{p}))$ embeds into the periodic group $GL(n, F)$), so $A^m \subseteq 1 + \mathbf{p}$ for some positive integer m and thus $A^m \subseteq 1 + \mathbf{p}_G$. Clearly $\mathbf{p}_G$ annihilates $(^*\mathbf{p})x$ for every x in G, so 9.6 yields that $\mathbf{p}_G$ annihilates $(^*\mathbf{p})R$, which equals M since M is irreducible. Consequently M can be considered as an $F(G/A^m)$-module and the induction hypothesis yields the desired result. $\square$

9.9   Lemma Let J be a commutative Noetherian Hilbert ring (for example the integers $\mathbf{Z}$), G a polycyclic-by-finite group and R the group ring JG. If M is an irreducible R-module, then M is annihilated by some maximal ideal of J.

P. Hall proved this for $J = \mathbf{Z}$ in 1959. To say the commutative ring J is Hilbert means that the Jacobson radical of each image K of J is equal to the intersection of all the prime ideals of K; that is, is equal to the prime radical of K. It is equivalent to each prime ideal of J being an intersection of maximal ideals of J.

Proof We have to show that $\mathbf{q} = \mathrm{Ann}_J M = J \cap M^*$ is a maximal ideal of J. We may pass to $J/\mathbf{q}$ and assume that $\mathbf{q} = \{0\}$. If $\alpha \in J$, then $^*\alpha$ is an R-submodule of M, since J is central. The R-irreducibility of M yields that $\alpha = 0$ or $^*\alpha = \{0\}$. Consequently M is J-torsion-free and J is a domain. Apply 9.3(b) with $A = \langle 1 \rangle$ and $S = J$. Hence there is a non-zero ideal $\mathbf{a}$ of J such that if $\mathbf{m}$ is a maximal ideal of J not containing $\mathbf{a}^g$ for any g in G, then $\mathbf{m}_G = \{0\}$. Here for $\mathbf{m}$ any maximal ideal of J we have $\mathbf{m}^g = \mathbf{m}$ for all $g \in G$, so if J is not a field, then $\mathbf{m}_G = \mathbf{m} \neq \{0\}$ and $\bigcap \mathbf{m} \geq \mathbf{a} \neq \{0\}$. But J is a Hilbert domain, so its Jacobson radical is $\{0\}$. This contradiction shows that J is a field. The result follows. $\square$

From 9.8 and 9.9 the following theorem is immediate. In many ways it is the main result of this chapter.

9.10   Theorem (Roseblade 1973a) Let J be a commutative Noetherian Hilbert ring all of whose field images are locally finite (for example the integers $\mathbf{Z}$). If G is a polycyclic-by-finite group, any irreducible JG-module is finite-dimensional over the appropriate field image of J.

9.11   Corollary (Roseblade 1973a) If G is a polycyclic-by-finite group, then every irreducible $\mathbf{Z}G$-module is finite.

This is immediate from 9.10, since the field images of $\mathbf{Z}$ are all finite. A module M is said to be *monolithic* if $M \neq \{0\}$ and if the intersection L of all the non-zero submodules of M is non-zero. L is then irreducible and is the only irreducible submodule of M. This submodule L is sometimes called the *lith* of M,

9.12   Corollary (Jategaonkar 1974; Roseblade 1976) If G is a polycyclic-by-finite group, then every finitely generated monolithic $\mathbf{Z}G$-module is finite.

Proof Let M be a finitely generated monolithic $\mathbf{Z}G$-module with lith L. By passing to $G/C_G(M)$ we may assume that G acts faithfully on M. By 9.11 the lith L is finite, so $C = C_G(L)$ has finite index in G. Let A be any abelian normal subgroup of G contained in C. Now L has finite prime exponent, p say. Set $\mathbf{b} = p\mathbf{Z}G + \mathbf{a}\mathbf{Z}G$, where $\mathbf{a}$ is the augmentation ideal of A. Clearly $\mathbf{b}$ is generated by elements of the centre of $\mathbf{Z}A$, so by 6.9 the ideal $\mathbf{b}$ is right strong Artin-Rees. Hence there is an integer $m > 0$ with

$$M\mathbf{b}^m \cap L \le L\mathbf{b} = \{0\}.$$

By definition of L we have $M\mathbf{b}^m = \{0\}$. Thus firstly $Mp^m = \{0\}$ and secondly A stabilizes the series

$$M \ge M\mathbf{a} \ge M\mathbf{a}^2 \ge \cdots \ge M\mathbf{a}^m = \{0\}.$$

By 1.21 (and our reduction to G acting faithfully on M) A is a p-group and hence is finite. This is for every such A. Thus C and therefore G are finite. Consequently M is finite.   □

9.13   Theorem   Finitely generated abelian-by-polycyclic-by-finite groups are residually finite.

This is the group-theoretic application we have used to motivate much of the preceding chapters. With polycyclic replaced by nilpotent, this was proved by P. Hall in 1959. At the same time he conjectured 9.13. Hall's case was far from easy, but 9.13 turned out to be very much harder indeed. However the necessary work has now been done and all we have left to do is essentially to repeat Hall's proof once he had reached this stage.

Proof Let E be a finitely generated group with an abelian normal subgroup A such that $G = E/A$ is polycyclic-by-finite. Let $a \in A \setminus \langle 1 \rangle$. Pick $B \le A$ with $a \notin B$, B normal in E and B maximal subject to these properties. Note that B exists by 3.10, the maximal condition on normal subgroups for such groups E; alternatively use Zorn's Lemma (1.11) if you prefer. Then A/B is a monolithic G-module and it is finitely generated by 3.9. Hence 9.12 yields that A/B is finite. But then E/B is polycyclic-by-finite and hence residually finite by 2.10. Consequently there exists a normal subgroup N of E of finite index with $B \le N$ and $a \notin N$. Also by 2.10 the group $E/A = G$ is residually finite. It follows that E is residually finite.   □

We state without proof some further results in this area. Using much the same techniques as above Segal has extended 9.13 as follows.

9.14   **Theorem** (Segal 1975b) Let E be a finitely generated group with an abelian normal subgroup A such that $G = E/A$ is polycyclic-by-finite. Then the following hold.

(a) There exists a finite set π of primes such that E is a finite extension of a group that is residually a finite nilpotent π-group.
(b) If A is torsion-free, then for almost all primes p, the group E is a finite extension of a group that is a residually finite p-group.
(c) If A is a p-group for some prime p, then E is a finite extension of a group that is a residually finite p-group.

See Segal (1975b) for a proof. Using very different techniques, in particular Krull dimension, Segal shows in Segal (1977) that a finitely generated $\mathbf{Z}G$-module over the finitely generated nilpotent-by-finite group G is poly (residually irreducible). He also proves the analogous result for certain other coefficient rings (specifically for commutative Noetherian Hilbert rings of Krull dimension at most 1 all of whose field images are locally finite). Again extending this to polycyclic groups turned out to be much more difficult than the original result. It was eventually settled by C.J.B. Brookes partially in his Cambridge Ph.D. thesis of 1981 and completed in 1988.

9.15   **Theorem** (Brookes 1988) Let J be a commutative Noetherian Hilbert ring with all of its field images locally finite, let G be a polycyclic-by-finite group and let M be a finitely generated JG-module. Then M is poly (residually irreducible).

Using similar techniques to those we have developed above, Roseblade has also proved the following two results. They follow easily from 9.15.

9.16   **Theorem** (Roseblade 1973a) Let J and G be as in 9.15. Then the Jacobson radical of $JG/\mathbf{a}$ is nilpotent for every ideal $\mathbf{a}$ of JG.

9.17   **Theorem** (Roseblade 1973a) Let J and G be as in 9.15. Let M be a finitely generated JG-module, H a normal subgroup of G and $\mathbf{h}$ its augmentation ideal in JH.

(a) If H is nilpotent and if $\mathbf{h}$ annihilates every JG-irreducible image of M, or
(b) if $\mathbf{h}$ annihilates every JG-chief factor of M,
    then some power of $\mathbf{h}$ annihilates M.

We conclude this chapter with a brief discussion of the primitive ideals of the group algebra of a polycyclic group. Recall that an ideal $\mathbf{a}$ of a ring R is (right) primitive if $\mathbf{a}$ is the annihilator in R of some irreducible right R-module M. Note that all primitive ideals of R are prime and that all maximal ideals are primitive (if $\mathbf{m}$ is a maximal ideal of R choose by Zorn's Lemma a maximal right ideal M of R containing $\mathbf{m}$ but not containing 1; then M is a maximal right ideal of R, R/M is irreducible and $\mathbf{m} = \mathrm{Ann}_R(R/M)$).

Let F be a field and G a polycyclic-by-finite group. Suppose F is locally finite and $\mathbf{p}$ is a primitive ideal of FG. Then by 9.8 and the Artin-Wedderburn Theorem, the ideal $\mathbf{p}$ of FG is maximal and $FG/\mathbf{p}$ is a matrix ring over a finite-dimensional division algebra. In particular primitive ideals of FG are incomparable (i.e. distinct ones cannot contain one another).

Suppose F is not locally finite. Then the situation is much more complex. Suppose we have a series

$$\langle 1 \rangle = H_0 < H_1 < H_2 < \cdots < H_r = H \le G,$$

where H has finite index in G, each H_i is normal in H, each H_i/H_{i-1} is free abelian and rationally irreducible as H-module and $r = pl(G)$, the plinth length of G (that is, r is maximal). Let $epl(G)$ be the number of the H_i/H_{i-1} of rank at least 2. This is the *eccentric plinth length* of G and 7.7 helps to explain why the dichotomy of rank 1 or rank at least 2 for a plinth can sometimes be significant. Notice that if G is finitely generated and nilpotent-by-finite then $epl(G) = 0$.

9.18 Theorem (Roseblade 1978) Let F be a field that is not locally finite and suppose G is a polycyclic-by-finite group.

(a) $epl(G)$ is the primitive length of FG; that is, $epl(G)$ is the maximal length of a chain of primitive ideals of FG.
(b) If $\mathbf{p}$ is a primitive ideal of FG, where $G \in \mathbf{Zr}$, then the primitive height of $\mathbf{p}$ in FG (the maximal length of an descending chain of primitive ideals starting at $\mathbf{p}$) is the plinth length of G in $\mathbf{p}^+ = G \cap (1 + \mathbf{p})$; that is the maximal length of that part of the series $\{H_i\}$ above lying in $\mathbf{p}^+$, over all possible choices of the series.

Modules over polycyclic groups have been studied for purely ring theoretic reasons, producing theorems that as yet do not seem to have much in the way of group theoretic implications. For example, Linnell et al. (2006) prove that all non-finitely generated projective modules over the polycyclic-by-finite group G are free if and only if G is polycyclic. The proofs here use Moody's induction theorem and also trace ideals. Brown (1981) studies the Krull dimension of $JG/\mathrm{Ann}_{JG} M$ for M a finitely generated module over JG with JG as in 9.10 above. It turns out to be the maximum of the Krull dimensions of the $JG/\mathrm{Ann}_{JG} N$ as N ranges over all the irreducible images of M. Musson (1981) studies the injective hulls of irreducible FG-modules for F a field and G a polycyclic-by-finite group.

Exercise Use 9.12 to prove that every finitely generated module over a polycyclic-by-finite group is residually finite. This also follows from 9.13; can you see why?

Chapter 10
Semilinear and Skew Linear Groups

In Chap. 4 we studied a polycyclic group by considering it as a subgroup of some $GL(n, \mathbf{Z})$. Is there some sort of analogous way of obtaining properties of finitely generated abelian-by-polycyclic groups? Firstly a consideration of linear groups will not suffice, since soluble linear groups must be nilpotent-by-abelian-by-finite. Indeed not even all finitely generated, abelian-by-polycyclic, nilpotent-by-abelian-by-finite groups (that is, in the notation of Chap. 1, $\mathbf{G} \cap \mathbf{AP} \cap \mathbf{NAF}$ groups) need be isomorphic to even quasi-linear groups (a *quasi-linear* group is a subgroup of a direct product of a finite number of linear groups; equivalently a quasi-linear group is any group of automorphisms of a Noetherian module over a commutative ring, see Wehrfritz 1979c, p. 55).

10.1 (Wehrfritz 1980a) Let $H = \langle x, y \rangle$ be a free nilpotent group of class 2 on 2 generators (so G is isomorphic to $Tr_1(3, \mathbf{Z})$). Set $z = [x, y]$ and $A = \mathbf{Z}H/(z-1)^2\mathbf{Z}H$. Let G be the split extension of A by H. Then

$$G \in \mathbf{G} \cap \mathbf{AN}_2 \cap \mathbf{N}_2\mathbf{A} \cap \mathbf{F}^{-S}$$

and yet G is not isomorphic to a group of automorphisms of any finitely generated module over a commutative ring.

($\mathbf{N}_2$ denotes the class of nilpotent groups of class at most 2 and $\mathbf{F}^{-S}$ denotes the class of torsion-free groups.) Thus to have any hope of success we need a hypothesis weaker than linearity. As we weaken the hypotheses we increase the number of groups we can faithfully represent, but increase the difficulty of proving something significant. Which reasonable possibilities do we have? Instead of automorphisms of finitely generated modules over fields we could consider one of the following.

(a) Automorphisms of non-finitely-generated modules over, for example, the integers or a field.
(b) Automorphisms of finitely generated modules over arbitrary commutative rings.
(c) Subgroups of $GL(n, D)$ for D now just a division ring rather than a field. Such subgroups are called *skew linear groups*.

B.A.F. Wehrfritz, *Group and Ring Theoretic Properties of Polycyclic Groups*, Algebra and Applications 10, DOI 10.1007/978-1-84882-941-1_10, © Springer-Verlag London Limited 2009

(d) Automorphisms of finitely generated modules over non-commutative rings.
(e) Automorphisms of some special but non-abelian group V, for example, V a finitely generated metabelian group. (This generalizes embedding our group into $GL(n, \mathbf{Z}) \cong \operatorname{Aut} V$, where V is a free abelian group of rank n.)
(f) Semilinear maps: Let M be a (right) module over a ring R. A *semilinear map* of M is an additive automorphism g of M for which there exists an automorphism γ of R such that $m\alpha g = mg\alpha^\gamma$ for all m in M and α in R. The set of all these forms a subgroup $\operatorname{Saut}_R M$ of $\operatorname{Aut}_{\mathbf{Z}} M$ (for if also $m\alpha h = mh\alpha^\eta$ for all m and α, then $m\alpha gh = mgh\alpha^{\gamma\eta}$ and if we set $m = ng^{-1}$ and $\beta = \alpha^\gamma$, then $ng^{-1}\alpha = n\alpha^\gamma g^{-1}$ and $n\beta g^{-1} = ng^{-1}\beta^\delta$ for $\delta = \gamma^{-1}$). Also $\operatorname{Aut}_R M \leq \operatorname{Saut}_R M$; for just choose the γ to be 1).

Call γ an *auxiliary automorphism* of g. Now γ is not necessarily unique for a given g. If γ' is a second auxiliary automorphism for g, then $mg\alpha^{\gamma'} = m\alpha g = mg\alpha^\gamma$ and $Mg = M$. Thus $\alpha^{\gamma'} - \alpha^\gamma$ lies in $\mathbf{a} = \operatorname{Ann}_R M$. Further if $\alpha \in \mathbf{a}$, then $mg\alpha^\gamma = m\alpha g = 0$, so $\alpha^\gamma \in \mathbf{a}$. Thus γ and γ' normalize $\mathbf{a}$ and define the same automorphism $\gamma^\wedge$ of $R^\wedge = R/\mathbf{a}$. If $\gamma^\wedge = 1$, then $m\alpha^\gamma = m\alpha$ for all m in M and so in this case g is linear. Therefore the sequence

$$1 \to \operatorname{Aut}_R M \to \operatorname{Saut}_R M \to \operatorname{Aut} R^\wedge$$

is exact, where the second arrow indicates containment and the third is the map $g \mapsto \gamma^\wedge$.

Suppose U is a group and M is a U-module. Then $g \in \operatorname{Aut}_{\mathbf{Z}} M$ is *semilinear* if there exists some γ in $\operatorname{Aut} U$ with $mug = mgu^\gamma$ for all m in M and u in U. Again these form a group $\operatorname{Saut}_U M$ with

$$\operatorname{Aut}_U M \leq \operatorname{Saut}_U M \leq \operatorname{Aut}_{\mathbf{Z}} M.$$

Set $C = C_U(M)$ and $U^\wedge = U/C$. If γ is an auxiliary automorphism of $g \in \operatorname{Saut}_U M$, then γ normalizes C and if γ' is a second such then $u^\gamma \equiv u^{\gamma'}$ modulo C. Thus γ defines a unique automorphism $\gamma^\wedge$ of $U^\wedge$ and we obtain an exact sequence

$$1 \to \operatorname{Aut}_U M \to \operatorname{Saut}_U M \to \operatorname{Aut} U^\wedge,$$

exactly as in the ring case.

Notice that in either case we obtain unique auxiliary automorphisms whenever the action (of R or U) on M is faithful. In practice it does not matter which case we take. If $U(R)$ is the group of units of R, then any automorphism of R normalizes $U(R)$ and so induces an automorphism of $U(R)$. Conversely any automorphism of our group U extends by linearity to an automorphism of the group ring $\mathbf{Z}U$. Thus

$$\operatorname{Saut}_R M \leq \operatorname{Saut}_{U(R)} M \quad \text{and} \quad \operatorname{Saut}_U M \leq \operatorname{Saut}_{\mathbf{Z}U} M.$$

(These two containments are not necessarily equalities.) We now comment on the five possibilities (a) to (e) above.

(a) This is far too wide, since any countable group can be embedded into $GL(\aleph_0, 2)$ for example.

(b) This is too restricted because of the following, see Wehrfritz (1979c), which effectively takes us back into the linear case.

10.2   Theorem If G is a finitely generated subgroup of $\mathrm{Aut}_R M$, where M is a finitely generated module over the commutative ring R, then there exists a finite collection of fields $F_1, F_2, \ldots, F_r$ and an integer n such that G can be embedded into

$$\underset{1 \le i \le r}{\times} GL(n, F_i) \cong GL\left(n, \bigoplus_{1 \le i \le r} F_i\right).$$

(c) and (d) As a promising start we have the following.

10.3   Theorem (Wehrfritz 1984b) Let E be a finitely generated group with an abelian normal subgroup A such that G/A is polycyclic-by-finite. Then E is isomorphic to a subgroup of the group of units of a semisimple Artinian ring of positive characteristic. Further, if the maximum periodic subgroup T of A is finite, then E can be embedded into $GL(n, D)$ for some integer n and division ring D of characteristic zero and if T is a finite extension of a p-group for some prime p, then E can be embedded into $GL(n, D)$ for some integer n and division ring D of characteristic p.

To exploit this we need to know about skew linear groups, that is about subgroups of $GL(n, D)$ for positive integers n and division rings D. We do know a fair amount about such subgroups, see Shirvani and Wehrfritz (1986), but not nearly as much as we know about linear groups and not nearly enough to be able to exploit the above theorem.

Quite a bit is known if the division ring D itself is generated over its centre by a polycyclic group; that is, if $D = F(H)$, where F denotes the centre of D and H is a polycyclic subgroup of D^*. These division algebras are a little more common than one might think, especially in the context of our discussion here. From a theorem of Brown and Wehrfritz (1984), see 4.1.1 of Shirvani and Wehrfritz (1986), if **p** is a prime ideal of the group ring JG, where J is any commutative ring and G is any polycyclic-by-finite group, then for some integer n and some division F-algebra $D = F(H)$ as above, the group of units of the ring of quotients of the ring $JG/\mathbf{p}$ embeds into $GL(n, D)$. For some properties of subgroups of $GL(n, D)$ for $D = F(H)$ as above, see Chap. 4 of Shirvani and Wehrfritz (1986), Wehrfritz (1991a) and Lichtman (1992); see also the end of Chap. 7 above.

As far as (d) explicitly is concerned apart from 10.2, from $R = D$ a division ring and from the case where R is a finite-dimensional algebra, I do not know anything useful in this context about subgroups of $\mathrm{Aut}_R M$ for potentially non-commutative rings R.

(e) and (f) Again we are off to a promising start in these cases.

10.4   Theorem (Wehrfritz 1975) Let G be a finitely generated, metabelian-by-finite group. Then there exists a finitely generated, abelian group U and a finitely generated U-module M such that the holomorph $\operatorname{Hol} G = G]\operatorname{Aut} G$ embeds into $\operatorname{Saut}_U M$. Conversely if U is a finitely generated, abelian group and if M is a finitely generated U-module, there exists a finitely generated, metabelian group G such that $\operatorname{Saut}_U M$ embeds into $\operatorname{Aut} G$.

In particular we have the simple result that if G is a finitely generated, metabelian-by-finite group, then there is a finitely generated, metabelian group H such that $\operatorname{Hol} G$ embeds into $\operatorname{Aut} H$. It also follows from 10.4 that in our context it suffices to focus on just one of (e) and (f). We choose (f).

For the remainder of this chapter U denotes some finitely generated abelian group and M a finitely generated U-module.

10.5   If G is a soluble subgroup of $\operatorname{Saut}_U M$, then G is nilpotent by polycyclic.

This is a good start, especially as it is not difficult to check that the wreath product $E = A \operatorname{wr} G$ embeds into some $\operatorname{Saut}_U M$ as above, where A is a finitely generated abelian group and G is polycyclic-by-finite (so $E \in \mathbf{G} \cap \mathbf{APF}$). To prove 10.5 use 4.4 and Wehrfritz (1973a) 13.26 & 13.29.

Conjecture If E is a finitely generated abelian-by-polycyclic-by-finite group, then there exists U and M as above such that E embeds into $\operatorname{Saut}_U M$.

Let G be any group and A a G-module. Say A satisfies (∗) as a G-module if there exists U and M as above and a homomorphism of G into $\operatorname{Saut}_U M$ (so M is now also a G-module) and a G-embedding of A into M. The next seven results can be found in Wehrfritz (1980a), with the exceptions of 10.9 and 10.12, which are unpublished but can be proved by very similar methods.

10.6   If for all finitely generated modules A over polycyclic-by-finite groups G, the G-module A has (∗), then every finitely generated abelian-by-polycyclic-by-finite group embeds into $\operatorname{Saut}_U M$ for some U and M as above.

10.7   It suffices to prove (∗) just for cyclic modules over polycyclic-by-finite groups.

10.8   If H is a subgroup of the polycyclic-by-finite group G, then $\mathbf{Z}G/\mathbf{h}\mathbf{Z}G$ has (∗) as G-module. (That is, cyclic permutation modules satisfy (∗); $\mathbf{h}$ here denotes the augmentation ideal of H.)

10.9   If G is a polycyclic-by-finite group and $\mathbf{a}$ is a semiprime ideal of the group ring $\mathbf{Z}G$, then $\mathbf{Z}G/\mathbf{a}$ has (∗) as G-module. (A semiprime ideal is an intersection of prime ideals.)

10.10   Let G be a polycyclic-by-finite group and p a prime. Then there exists a normal subgroup H of G of finite index such that for every ideal $\mathbf{a}$ of the group algebra $\mathbf{F}_p H$, the factor $\mathbf{F}_p H/\mathbf{a}$ has $(*)$ as H-module.

Thus the evidence for the truth of the $(*)$ conjecture is promising and hence it seems quite likely that finitely generated abelian-by-polycyclic-by-finite groups do embed into suitable Saut$_U$ M. Conversely we have the following.

10.11   If E is an abelian-by-polycyclic-by-finite subgroup of Saut$_U$ M, then there exists an abelian normal subgroup A of E such that $G = E/A$ is polycyclic-by-finite and A has poly $(*)$ as G-module.

Thus perhaps $(*)$ is too strong and one can only prove poly $(*)$. The following at least holds.

10.12   If G is a polycyclic-by-finite group and if $\mathbf{a}$ is an ideal of $\mathbf{Z}G$, then $\mathbf{Z}G/\mathbf{a}$ satisfies poly $(*)$.

Suppose the conjecture has been proved. What does one know about these groups Saut$_U$ M of semilinear maps that would give useful information about finitely generated abelian-by-polycyclic-by-finite groups? Naturally we are especially interested in the soluble subgroups of Saut$_U$ M. Below we wish to compare properties of subgroups of Saut$_U$ M with known or suspected properties of $(\mathbf{G} \cap \mathbf{APF})$ groups. For an account of properties of $(\mathbf{G} \cap \mathbf{APF})$ groups and especially $(\mathbf{G} \cap \mathbf{ANF})$ groups, see Chap. 7 of Lennox and Robinson (2004).

10.13   Saut$_U$ M is residually finite. More generally there exist finite sets π and κ of primes such that for all primes $p \notin \kappa$ the group Saut$_U$ M is a finite extension of a residually (finite nilpotent $\pi \cup \{p\}$-group).

Given the truth of the $(*)$ conjecture, this would imply, for example, that irreducible modules over polycyclic-by-finite groups are finite and that $(\mathbf{G} \cap \mathbf{APF})$ groups are residually finite, facts that earlier required from us much work, see 9.11 and 9.13. It would also yield 9.14(a) that we left without proof. Yet 10.13 is very easy. The following result, however, is certainly not very easy.

10.14   (Wehrfritz 1979b) Given our U and M there exists a positive integer z such that if $G \le$ Saut$_U$ M and if Z is a subset of G such that for every $x \in Z$, $g \in G$ and characteristic ideal $\mathbf{a}$ of $\mathbf{Z}U$ of finite index, there is a positive integer n with

$$[x,_n g] = [x, g, g, \ldots, g] \in C_G(M/M\mathbf{a}),$$

then $[\langle Z \rangle,_z G] = \langle 1 \rangle$; equivalently $Z \subseteq \zeta_z(G)$.

This theorem has many corollaries, see Wehrfritz (1979b). In 10.15 the integer z is as in 10.14.

10.15 (a) $\mathrm{Saut}_U\, M$ is centrally stunted; that is, there is a bound on the central heights of the subgroups of $\mathrm{Saut}_U\, M$.

In fact these heights are at most z. This is not known yet as far as I know for $(\mathbf{G} \cap \mathbf{APF})$ groups, but is known for $(\mathbf{G} \cap \mathbf{ANF})$ groups, see Lennox and Roseblade (1970).

(b) If G is a subgroup of $\mathrm{Saut}_U\, M$, then the Frattini subgroup $\Phi(G)$ is nilpotent and $\eta(G/\Phi(G)) = \eta(G)/\Phi(G)$ and is the unique maximal abelian normal subgroup of $G/\Phi(G)$. Further $\delta(G)$, the intersection of the non-normal maximal subgroups of G (meaning G if none such exist), is also nilpotent and $\delta(G)/\Phi(G)$ is the centre of $G/\Phi(G)$.

This is already known for $(\mathbf{G} \cap \mathbf{APF})$ groups, see Roseblade (1973a).

(c) For any subgroup G of $\mathrm{Saut}_U\, M$, the Hirsch Plotkin radical $\eta(G)$ is nilpotent of class at most z and is equal to both the set of left Engel elements of G and the set of bounded left Engel elements of G. Also $\zeta(G) = \zeta_z(G)$ and is equal to both the set of right Engel elements of G and the set of bounded right Engel elements of G.

These properties are known to hold in $(\mathbf{G} \cap \mathbf{APF})$ groups (Gruenberg (1961) Theorem 1.5 coupled with our comments on (a) above).

(d) If G is any subgroup of $\mathrm{Saut}_U\, M$, then $\eta(G) = \psi(G) = \psi_1(G)$, where $\psi(G)$ denotes the intersection of the centralizers in G of all the chief factors of G and $\psi_1(G)$ denotes the intersection of the centralizers in G of all the finite chief factors of G.

These equalities are known to hold for $(\mathbf{G} \cap \mathbf{APF})$ groups, see Roseblade (1973a).

(e) Let H and K be subgroups of $\mathrm{Saut}_U\, M$ such that for all characteristic ideals $\mathbf{a}$ of $\mathbf{Z}G$ of finite index there exists an integer m with $[H, {}_m K, M] \le M\mathbf{a}$. Then $[H, {}_{z+1} K] = \langle 1 \rangle$.

An analogue of this is not known to hold for $(\mathbf{G} \cap \mathbf{APF})$ groups but does hold for $(\mathbf{G} \cap \mathbf{ANF})$ groups, see Segal (1977).

10.16 (Wehrfritz 1978a) $\mathrm{Saut}_U\, M$ is centrally eremitic; that is, there exists an integer e such that all elements x and y of $\mathrm{Saut}_U\, M$ for which there exists a positive integer n with $[x^n, y] = 1$, satisfy $[x^e, y] = 1$.

The condition $e = 1$ is equivalent to centralizers being isolated. Again this is not known to hold for $(\mathbf{G} \cap \mathbf{APF})$ groups but does hold for $(\mathbf{G} \cap \mathbf{ANF})$ groups, see Lennox and Roseblade (1970). It is also known for torsion-free $(\mathbf{G} \cap \mathbf{APF})$ groups, see Segal (1975b).

The *centralizer dimension* (or central gap number) of a group G is the length of the longest chain of distinct centralizers (of subsets) in G, assuming these lengths are bounded, and ∞ otherwise. Probably $\mathrm{Saut}_U\, M$ has finite centralizer dimension, but I cannot prove it. However we are primarily interested in soluble subgroups of $\mathrm{Saut}_U\, M$ and these do have finite centralizer dimension. More generally we have the following.

10.17 (Wehrfritz 1978a) Given our U and M there exists an integer d such that if G is any subgroup of $\mathrm{Saut}_U\, M$ with $G/(G \cap \mathrm{Aut}_U\, M)$ soluble-by-finite, then G has finite centralizer dimension at most d.

Once again it does not seem to be known whether all $(\mathbf{G} \cap \mathbf{APF})$ groups have finite centralizer dimension, but $(\mathbf{G} \cap \mathbf{ANF})$ groups do, see Lennox and Roseblade (1970).

The *central product rank* of a group G is the least upper bound of all integers n such that G contains subgroups $G_1, G_2, \ldots, G_n$ with $[G_i, G_j] = \langle 1 \rangle$ if and only if $i \neq j$. The centralizer dimension of a group always bounds its central product rank.

10.18 (Wehrfritz 1978a) Saut$_U$ M has finite central product rank.
This is not I think known for $(\mathbf{G} \cap \mathbf{APF})$ groups.

A final thought; the techniques used to prove some of the results of this chapter (especially 10.7 to 10.12, see Wehrfritz 1980a) can be used in a completely different context. Let G be a polycyclic-by-finite group and let H be a subgroup of G. By 4.8 we can embed G into some $GL(n, \mathbf{Z})$. It is shown in Wehrfritz (1980a) that we can do this in such a way that the image of H is (Zariski) closed in the image of G, or indeed is even the fixed-point set in G of some vector (see 2.1 and 2.6 of Wehrfritz 1980a). This is some of the extra information needed for the results above of Chap. 10 and for what comes below.

If H is a subgroup of a group G and if Γ is a group acting on G, let $\mathbf{L}(\Gamma, H)$ denote the poset, ordered by inclusion, of all possible intersections $\bigcap_{\gamma \in \Sigma} H^\gamma$ as Σ ranges over all subsets of Γ. An important special case is where $\Gamma = G$ with the action being conjugation.

Continuing the above notation, assume G is also polycyclic-by-finite. Trivially $\mathbf{L}(G, H)$ satisfies the maximal condition, since G satisfies the maximal condition on subgroups. A surprising theorem of Rhemtulla (1967) is that the $\mathbf{L}(G, H)$ also satisfies the minimal condition. His proof is far from trivial but essentially uses the techniques of our Chap. 2. However this theorem is immediate from our remark above, since linear groups satisfy the minimal condition on closed subgroups and if $H \leq G \leq GL(n, \mathbf{Z})$ is closed in G, so is every member of $\mathbf{L}(G, H)$. More work, but using the same general ideas, proves the more general result that $\mathbf{L}(\Gamma, H)$ satisfies the minimal condition for any subgroup H of a polycyclic-by-finite group G with Γ any group of automorphisms of G, see Wehrfritz (2009b). Indeed chains in such an $\mathbf{L}(\Gamma, H)$ have bounded length.

Conversely it follows from 3.10 of Chap. 3 that any soluble-by-finite group G in which $\mathbf{L}(G, H)$ satisfies the minimal condition for each subgroup H of G, necessarily is polycyclic-by-finite, see Rhemtulla (1967).

Again let H be a subgroup of a group G. It is natural to ask about the analogous poset $\mathbf{U}(G, H)$ of all subgroups $\langle H^g : g \in S \rangle$ as S ranges over all subsets of G, again ordered by inclusion. It turns out in this case that it suffices for most purposes just to consider cyclic subgroups H. Say the group G satisfies max C if $\mathbf{U}(G, \langle x \rangle)$ satisfies the maximal condition for all elements x of G. For example, clearly all FC-groups satisfy max C. Then a soluble-by-finite group satisfies max C if and only if it is polycyclic-by-finite. An $\langle L, P \rangle(\mathbf{AF})$-group satisfies max C if and only if it is generated by its polycyclic-by-finite normal subgroups. If G is a linear group the following are equivalent.

(a) G satisfies max C.
(b) G' is polycyclic-by-finite.
(c) $G/\zeta_1(G)$ is polycyclic-by-finite.

All these results on the condition max C can be found in Wehrfritz (2009c), where the obvious analogous concept of min C is also discussed.

Notation

Abbreviations

max: the maximal condition on subgroups, see p. 16

max-n: the maximal condition on normal subgroups, see p. 37

min: the minimal condition of subgroups, see p. 38

Group Notation

Below G is a group, H a subgroup of G, N a normal subgroup of G, X and Y subsets of G, x and y elements of G and K a group acting (as automorphisms) on G

Conjugates: $x^y = x^{-1}yx$ and $x^G = \{x^y : y \in G\}$

Commutators: $[x, y] = x^{-1}y^{-1}xy$ and $[X, Y] = \langle [x, y]x \in X \& y \in Y \rangle$

$N_G(X)$: the normalizer of X in G

$C_G(X)$: the centralizer of X in G

$G', G'', G^{(n)}$: derived subgroups, see p. 3

$\{\gamma^n G\}$: the lower central series of G, see p. 3

$\{\zeta_n(G)\}$: the upper central series of G, see pp. 3 and 63

$\zeta_1(G)$: the centre of G

$\zeta(G)$: the hypercentre of G

Fitt $G = \eta_1(G)$: the Fitting subgroup of G, see p. 4

$\Phi(G)$: the Frattini subgroup of G, see p. 6

$G = N]H = H[N$: split extension/semi-direct product, see p. 2

G_0: see pp. 43 and 44

B.A.F. Wehrfritz, *Group and Ring Theoretic Properties of Polycyclic Groups*,
Algebra and Applications 10,
DOI 10.1007/978-1-84882-941-1, © Springer-Verlag London Limited 2009

G^0: see pp. 43 and 52

$u(G)$: the unipotent radical of G, see p. 49

$\tau(G)$: the unique maximal periodic normal subgroup of G, see p. 22

$H^G = \langle H^g : g \in G \rangle$

$H_G = \bigcap_{g \in G} H^g$

$i_G(H)$: the isolator of H in G, see p. 58

$i_G^\pi(H)$: the π-isolator of H in G. see p. 59

$\Delta_G(K) = \{g \in G : (K : C_K(g)) = |g^K| \text{ is finite}\}$, see p. 66

$\Delta(G) = \Delta_G(G)$

$\delta_G(H)$: see p. 94

Group Notation—Group Classes

A the class of abelian groups

C$_n$: the class of cyclic groups of order n or 1

F: the class of finite groups

F$_{\{p\}}$, **F**$_\pi$: the class of finite p-groups, resp. of finite π-groups

F^{-S}; the class torsion-free groups

G: the class of finitely generated groups

G$_1$: the class of cyclic ($=$ 1-generator) groups

I: the class of trivial groups $\langle 1 \rangle$

N: the class of nilpotent groups

N$_2$: the class of nilpotent groups of class at most 2

P: the class of polycyclic groups, see pp. 14 and 15

S: the class of soluble groups

U: the class of all groups

Max: the class of groups satisfying the maximal condition on subgroups, see p. 16

Max-n: the class of groups satisfying the maximal condition on normal subgroups, see p. 38

Min: the class of groups satisfying the minimal condition of subgroups, see p. 38

Operators on Classes

L: the local operator, see p. 15

P: the poly operator, see p. 14

Q: the quotient operator, see p. 13

R: the residual operator, see p. 15

S: the subgroup operator, see p. 13

Rings and Modules

Z: the integers

Q: the rational numbers

R: the real numbers

C: the complex numbers

F$_p$: the field of order p

Z$_p$: the p-adic integers

Q$_p$: the p-adic numbers

F^*: the multiplicative group of the field F

$\text{Ann}_R M$: the annihilator of the module M in the ring R

$S[G]$: a ring generated by the subgroup G of its group of units and its subring S normalized by G, see p. 68

$\mathbf{p}^+$: see p. 82

$^*X$ and Y^*: see p. 102

$\pi_R M$: see p. 102

Special Functions

$h(G)$: the Hirsch number of G, see p. 18

$pl(G)$: the plinth length of G, see p. 97

$ht(\mathbf{p})$: the height of the prime ideal $\mathbf{p}$, see p. 97

$[r]$: the greatest integer not exceeding the real number r

Special Groups

$\text{Aut}_R M$, $\text{Aut}_G M$: the automorphism group of M as R-module, resp. as G-module

$\text{Aut}_{\mathbf{Z}} M$: the automorphism group of the additive group of the module M

$GL(n, F)$, $GL(n, R)$: the general linear group, see p. 41

$D(n, F)$: the diagonal group, see p. 41

$Tr(n, F)$: the (lower) triangular group, see p. 41

$Tr_1(n, F)$, $Tr^1(n, F)$: the lower, resp. upper, unitriangular group, see p. 41

$\text{Saut}_R M$, $\text{Saut}_G M$: see p. 110

Special Symbols

$\leq$: is a subgroup of

$<$: is a subgroup of but not the whole group

$\lhd$: is a normal subgroup of

$\lhd\lhd$: is a subnormal subgroup of

$\times$, $\oplus$: direct product and sum signs, see p. 2

Bibliography

Almazar VD, Cossey J (1996) Polycyclic products of nilpotent groups. Arch Math (Basel) 66:1–7

Auslander L (1967) On a problem of Philip Hall. Ann Math (2) 86:112–116

Auslander L (1969) The automorphism group of a polycyclic group. Ann Math (2) 89:314–322

Baer R (1955a) Nilgruppen. Math Z 62:402–437

Baer R (1955b) Auflösbare Gruppen mit Maximalbedingung. Math Ann 129:139–173

Baer R (1957) Überauflösbare Gruppen. Abh Math, Sem Univ Hamb 23:11–28

Baues O, Grunewald F (2006) Automorphism groups of polycyclic-by-finite groups and arithmetic groups. Publ Math Inst Hautes Études Sci 104:213–268

Baumslag G (1974) Residually finite groups with the same finite images. Compos Math 29:249–252

Baumslag G, Cannonito FB, Robinson DJS, Segal D (1991) The algorithmic theory of polycyclic-by-finite groups. J Algebra 142:118–149

Bergman GM (1971) The logarithmic limit set of an algebraic variety. Trans Am Math Soc 157:459–469

Blackburn N (1965) Conjugacy in nilpotent groups. Proc Am Math Soc 16:143–148

Brewster DC (1976) The maximum condition on ideals of a group ring. PhD thesis. Cambridge University

Brookes CJB (1988) Modules over polycyclic groups. Proc Lond Math Soc (3) 57:88–108

Brown KA (1981) Modules over polycyclic groups have many irreducible images. Glasg Math J 22:141–150

Brown KA, Wehrfritz BAF (1984) Division rings associated with polycyclic groups. J Lond Math Soc (2) 30:465–467

Chatters AW, Hajarnavis CR (1980) Rings with chain conditions. Research notes in mathematics, vol 44. Pitman, London

Chevalley C (1951) Deux théorèmes d'arithmétique. J Math Soc Jpn 3:36–44

Cohn PM (1974) Algebra. Wiley, Chichester. 2nd edition, 1989

Cossey J (1991) The Wielandt subgroup of a polycyclic group. Glasg Math J 33:231–234

du Sautoy M (2002) Polycyclic groups, analytic groups and algebraic groups. Proc Lond Math Soc (3) 85:62–92

Endimioni G (1998) On the nilpotent length of polycyclic groups. J Algebra 203:125–133

Farkas DR (1982) Endomorphisms of polycyclic groups. Math Z 181:567–574

Fitting H (1938) Beiträge zur Theorie der Gruppen endlicher Ordnung. Jahresber Deutsch Math Verein 48:77–141

Formanek E (1970) Matrix techniques in polycyclic groups. PhD thesis, Rice University of Houston, TX

Formanek E (1976) Conjugate separability of polycyclic groups. J Algebra 42:1–10

Frattini G (1885a) Intorno alle generazione dei gruppi di operazioni I. Rend Accad Naz Lincei (4) 1:281–285

Frattini G (1885b) Intorno alle generazione dei gruppi di operazioni II. Rend Accad Naz Lincei (4) 1:455–457

Gruenberg KW (1957) Residual properties of infinite soluble groups. Proc Lond Math Soc (3) 7:29–62

Gruenberg KW (1961) The upper central series in soluble groups. Illinois J Math 3:436–466

Gruenberg KW (1973) Ring theoretic methods and finiteness conditions in infinite soluble groups. In: Proceedings of the conference on group theory 1972. Lecture notes in mathematics, vol 319. Springer, Berlin, pp 75–84

Grunewald FJ, Segal D (1978) Conjugacy in polycyclic groups. Commun Algebra 6:775–798

Grunewald FJ, Pickel PF, Segal D (1980) Polycyclic groups with isomorphic finite quotients. Ann Math (2) 111:155–195

Hall P (1954) Finiteness conditions for soluble groups. Proc Lond Math Soc (3) 4:419–436

Hall P (1957) Nilpotent groups. In: Lectures at 1957 Canadian mathematical congress. Re-issued as 'The Edmonton notes on nilpotent groups'. Mathematical congress, college math notes, London, 1969

Hall P (1958) Some sufficient conditions for a group to be nilpotent. Illinois J Math 2:787–801

Hall P (1959) On the finiteness of certain soluble groups. Proc Lond Math Soc (3) 16:595–622

Harper DL (1980) Primitivity in representations of polycyclic groups. Math Proc Camb Philos Soc 88:15–31

Higman G (1955) A remark on finitely generated nilpotent groups. Proc Am Math Soc 6:284–285

Hirsch KA (1937) On the class of infinite soluble groups. PhD thesis, Cambridge University

Hirsch KA (1938a) On infinite soluble groups I. Proc Lond Math Soc (2) 44:53–60

Hirsch KA (1938b) On infinite soluble groups II. Proc Lond Math Soc (2) 44:336–344

Hirsch KA (1946) On infinite soluble groups III. Proc Lond Math Soc (2) 49:184–194

Hirsch KA (1952) On infinite soluble groups IV. J Lond Math Soc 27:81–85

Hirsch KA (1954) On infinite soluble groups V. J London Math Soc 29:250–251

Holt D, Eick B, O'Brien E (2005) Handbook of computational group theory. Chapman & Hall/CRC Press, London

Hurley TC (1990) On the class of the stability group of a series of subgroups. J Lond Math Soc (2) 41:33–41

Ito N (1953) Note on S-groups. Proc Jpn Acad 29:149–150

Jacobson N (1985) Basic algebra, 2 vols. Freeman, New York

Jategaonkar AV (1974) Integral group rings of polycyclic-by-finite groups. J Pure Appl Algebra 4:337–343

Jeanes SC, Wilson JS (1978) On finitely generated groups with many profinite-closed subgroups. Arch Math (Basel) 31:120–122

Kaluzhnin LA (1950) Sur quelques propriétés des groupes d'automorphisme d'un groupe abstrait. C R Acad Sci Paris 230:2067–2069

Kaluzhnin LA, Krasner M (1951) Produit complete des groupes de permutations et le problème d'extension des groupes III. Acta Math Szeged 14:69–82

Kaplansky I (1970) Commutative rings. Allyn & Bacon, Boston

Kegel OH (1966) Über den Normalisator von subnormalen und erreichbaren Untergruppen. Math Ann 163:248–258

Kegel OH, Wehrfritz BAF (1973) Locally finite groups. North-Holland, Amsterdam

Knight JT (1971) Commutative algebra. Cambridge University Press, Cambridge

Kolchin ER (1948) Algebraic matrix groups and the Picard-Vessiot theory of homogeneous linear ordinary differential equations. Ann Math (2) 49:1–42

Learner A (1962) The embedding of a class of polycyclic groups. Proc Lond Math Soc (3) 12:496–510

Learner A (1964) Residual properties of polycyclic groups. Illinois J Math 8:536–542

Lennox JC (1976) Finitely generated metabelian groups are not subnormality separable. Math Z 149:201–202

Lennox JC, Robinson DJS (2004) The theory of infinite soluble groups. Clarendon, Oxford

Lennox JC, Roseblade JE (1970) Centrality in finitely generated soluble groups. J Algebra 16:399–435

Lennox JC, Roseblade JE (1980) Soluble products of polycyclic groups. Math Z 170:153–154

Lennox JC, Wilson JS (1977) A note on permutable subgroups. Arch Math (Basel) 28:113–116

Lennox JC, Wilson JS (1979) On products of subgroups in polycyclic groups. Arch Math (Basel) 33:305–309

Letzter ES, Lorenz M, (1999) Polycyclic-by-finite group algebras are catenary. Math Res Lett 6:183–194

Lichtman AI (1992) Trace functions in the ring of fractions of polycyclic group rings. Trans Am Math Soc 330:769–781

Linnell PA, Puninski G, Smith P (2006) Idempotent ideals and non-finitely generated projective modules over integral group rings of polycyclic-by-finite groups. J Algebra 305:845–858

Linnell PA, Warhurst D (1981) Bounding the number of generators of a polycyclic group. Arch Math (Basel) 37:7–17

Lorenz M, Passman DS (1981) Prime ideals in group algebras of polycyclic-by-finite groups. Proc Lond Math Soc (3) 43:520–543

Mal'cev AI (1940) On faithful representations of infinite groups of matrices. Mat Sb 8:405–422 (in Russian); Am Math Soc Transl (2) 45:1–18 (1965)

Mal'cev AI (1951) On certain classes of infinite soluble groups. Mat Sb 28:567–588 (in Russian); Am Math Soc Transl (2) 2:1–21 (1956)

Mal'cev AI (1958) Homomorphisms onto finite groups. Ivanov Gos Ped Inst Ucen Zap 18:49–60 (in Russian)

Mann A, Segal D (2007) Breadth in polycyclic groups. Int J Algebra Comput 17:1073–1083

McConnell JC (1968) Localisation in enveloping rings. J Lond Math Soc 43:421–428

McConnell JC, Robson JC (1987) Noncommutative Noetherian rings. Wiley, Chichester

Mennicke JL (1965) Finite factor groups of the unimodular group. Ann Math 81:316–337

Merzljakov JuI (1969) Matrix representations of groups of outer automorphisms of Chernikov groups. Algebra Log 8:478–482 (in Russian)

Merzljakov JuI (1970) Integral representations of the holomorph of a polycyclic group. Algebra Log 9:539–558 (in Russian)

Moravec P (2007) The non-abelian tensor product of polycyclic groups is polycyclic. J Group Theory 10:795–798

Musson IM (1981) Irreducible modules for polycyclic group algebras. Can J Math 33:901–914

Neumann BH (1954) Groups covered by permutable subsets. J Lond Math Soc 29:236–248

Nikolov N, Segal D (2007) Direct products and profinite completions. J Group Theory 10:789–793

Nouazé Y, Gabriel P (1967) Idéaux premiers de l'algèbre enveloppante d'une algèbre de Lie nilpotente. J Algebra 6:77–99

Ol'shanskii AYu (1979) Infinite groups with cyclic subgroups. Dokl Akad Nauk SSSR 245:785–789 (in Russian)

Ol'shanskii AYu (1982) Groups of bounded period with subgroups of prime order. Algebra Log 21:553–618

Ol'shanskii AYu (1991) Geometry of defining relations in groups. Kluwer Academic, Dordrecht

Passman DS (1977) The algebraic structure of group rings. Wiley, New York

Passman DS (1984) Group rings of polycyclic groups. In: Gruenberg KW, Roseblade JE (eds) Group theory essays for Philip Hall. Academic Press, London

Platonov VP (1966) The Frattini subgroups of linear groups and finite approximability. Dokl Akad Nauk SSSR 171:798–801 (in Russian); Sov. Math Dokl 7 (1966):1557-1560

Remeslennikov VN (1969) Representation of finitely generated metabelian groups by matrices. Algebra Log 8:72–75 (in Russian)

Rhemtulla AH (1967) A minimality property of polycyclic groups. J Lond Math Soc 42:456–462

Rhemtulla AH, Wehrfritz BAF (1984) Isolators in soluble groups of finite rank. Rocky Mt J Math 14:415–421

Rhemtulla AH, Wilson JS (1988) Elliptically embedded subgroups of polycyclic groups. Proc Am Math Soc 102:230–234

Ribes L, Segal D, Zalesskii PA (1998) Conjugacy separability and free products with cyclic amalgamations. J Lond Math Soc (2) 57:609–628

Robinson DJS (1970) A theorem on finitely generated hyperabelian groups. Invent Math 10:38–43

Robinson DJS (1972) Finiteness conditions and generalized soluble groups, 2 vols. Springer, Berlin

Robinson DJS (1980) A course in the theory of groups. Springer, Berlin

Robinson DJS (2002) Derivations and the permutability of subgroups in polycyclic-by-finite groups. Proc Am Math Soc 130:3461–3464

Roseblade JE (1971) The integral group rings of hypercentral groups. Bull Lond Math Soc 3:351–355

Roseblade JE (1973a) Group rings of polycyclic groups. J Pure and Appl Algebra 3:307–328

Roseblade JE (1973b) Polycyclic group rings and the Nullstellensatz. In: Proceedings of the conference on group theory, 1972. Lecture notes in mathematics, vol 319. Springer, Berlin, pp 156–167

Roseblade JE (1973c) The Frattini subgroup in infinite soluble groups. In: Three lectures on polycyclic groups. Queen Mary college math notes, London

Roseblade JE (1976) Applications of the Artin-Rees lemma to group rings, 1973 convegno sui gruppi infiniti. Sympos Math 17:471–478

Roseblade JE (1978) Prime ideals in group: rings of polycyclic groups. Proc Lond Math Soc (3) 36:385–447; Corrigenda: Proc Lond Math Soc (3) 38:216–218 (1979)

Roseblade JE, Smith PF (1976) A note on hypercentral group rings. J Lond Math Soc (2) 13:183–190

Roseblade JE, Smith PF (1979) A note on the Artin-Rees property of certain polycyclic group algebras. Bull Lond Math Soc 11:184–185

Samuel P (1972) Algebraic theory of numbers. Kershaw, London

Schmidt FK (1930) Zur Klassenkörpertheorie im Kleinen. J Reine Angew Math 162:155–166

Schur I (1904) Über die Darstellungen der endlichen Gruppen durch gebrochene lineare Substitutionen. J Reine Angew Math 127:20–50

Segal D (1975a) Groups whose finite quotients are supersoluble. J Algebra 35:56–71

Segal D (1975b) On abelian-by-polycyclic groups. J Lond Math Soc (2) 11:445–452

Segal D (1977) On the residual simplicity of certain modules. Proc Lond Math Soc (3) 34:327–353

Segal D (1978) Two theorems on polycyclic groups. Math Z 164:185–187

Segal D (1983) Polycyclic groups. Cambridge University Press, Cambridge

Segal D (1987) The general polycyclic group. Bull London Math Soc 19:49–56

Segal D (1990) Decidable properties of polycyclic groups. Proc Lond Math Soc (3) 61:497–528

Segal D (2000) On modules of finite upper rank. Trans Am Math Soc 353:391–410

Segal D (2001) On the group rings of abelian minimax groups. J Algebra 237:64–94

Segal D (2006) On the group rings of abelian minimax groups II: the singular case. J Algebra 306:379–396

Seksenbaev K (1965) On the theory of polycyclic groups. Algebra Log 4:79–83 (in Russian)

Shirvani M, Wehrfritz BAF (1986) Skew linear groups. Cambridge University Press, Cambridge

Shmel'kin AL (1968) Polycyclic groups. Sib Mat Zh 9:234–235; Sib Math J 9 (1968):178 (in Russian)

Sims CC (1994) Computation with Finitely Presented Groups. Cambridge University Press, Cambridge

Smirnov DM (1953) On groups of automorphisms of soluble groups. Mat Sb 32:365–384 (in Russian)

Stewart AGR (1966) On the class of certain nilpotent groups. Proc R Soc Ser A 292:374–379

Swan RG (1967) Representations of polycyclic groups. Proc Am Math Soc 18:573–574

Wang H-C (1956) Discrete subgroups of solvable Lie groups. Ann Math (2) 64:1–19

Wehrfritz BAF (1968) Frattini subgroups in finitely generated linear groups. J Lond Math Soc 43:619–622

Wehrfritz BAF (1970) Groups of automorphisms of soluble groups. Proc Lond Math Soc (3) 20:101–122

Wehrfritz BAF (1972) A note on residual properties of nilpotent groups. J Lond Math Soc (2) 5:1–7

Wehrfritz BAF (1973a) Infinite linear groups. Springer, Berlin

Wehrfritz BAF (1973b) The holomorph of a polycyclic group. In: Three lectures on polycyclic groups. Queen Mary college math notes, London

Wehrfritz BAF (1973c) Two examples of soluble groups that are not conjugacy separable. J Lond Math Soc (2) 7:312–316

Wehrfritz BAF (1974) On the holomorphs of soluble groups of finite rank. J Pure Appl Algebra 4:55–69

Wehrfritz BAF (1975) Representations of holomorphs of group extensions with abelian kernels. Math Proc Camb Philos Soc 78:357–367

Wehrfritz BAF (1976) Finitely generated groups of module automorphisms and finitely generated metabelian groups. Sympos Math 17:261–275

Wehrfritz BAF (1977) Nilpotence in groups of semi-linear maps II: a normalization theorem. J Lond Math Soc (2) 16:449–457

Wehrfritz BAF (1978a) The centralizer poset in groups of semilinear maps. Mathematika 25:251–263

Wehrfritz BAF (1978b) On the Lie-Kolchin-Mal'cev theorem. J Austral Math Soc (A) 26:270–276

Wehrfritz BAF (1979a) Invariant maximal ideals of commutative rings. J Algebra 59:472–480

Wehrfritz BAF (1979b) Nilpotence in groups of semi-linear maps III. J Pure Appl Algebra 15:93–107

Wehrfritz BAF (1979c) Lectures around complete local rings. Queen mary college math notes, London

Wehrfritz BAF (1980a) Finitely generated modules over polycyclic groups. Q J Math (2) 31:109–127

Wehrfritz BAF (1980b) On finitely generated soluble linear groups. Math Z 170:155–167

Wehrfritz BAF (1983) Endomorphisms of polycyclic groups. Math Z 184:97–99

Wehrfritz BAF (1984a) On division rings generated by polycyclic groups. Israel J Math 47:154–164

Wehrfritz BAF (1984b) Faithful representations of finitely generated abelian-by-polycyclic groups over division rings. Q J Math 35:361–372

Wehrfritz BAF (1991a) Polycyclic group algebras and theorems of Harper and Lichtman. Arch Math (Basel) 57:228–237

Wehrfritz BAF (1991b) On rings of quotients of group algebras of soluble groups of finite rank. J Pure Appl Algebra 74:95–107

Wehrfritz BAF (1992) Invariant maximal ideals in certain group algebras. J Lond Math Soc (2) 46:101–110

Wehrfritz BAF (1994) Two remarks on polycyclic groups. Bull Lond Math Soc 26:543–548

Wehrfritz BAF (1999) Finite groups. World Scientific, Singapore

Wehrfritz BAF (2009a) Endomorphisms of polycyclic-by-finite groups. Math Z (to appear)

Wehrfritz BAF (2009b) On a theorem of Rhemtulla on polycyclic groups. Preprint

Wehrfritz BAF (2009c) Variations on the theme of FC-groups. Ric Mat (to appear)

Wilson JS (1982a) Abelian subgroups of polycyclic groups. J Reine Angew Math 331:162–180

Wilson JS (1982b) Large nilpotent subgroups of polycyclic groups. Arch Math (Basel) 39:1–4

Zariski O, Samuel P (1958) Commutative algebra, vol 1. Van Nostrand, Princeton

Zariski O, Samuel P (1960) Commutative algebra, vol 2. Van Nostrand, Princeton

Index

B.A.F. Wehrfritz, *Group and Ring Theoretic Properties of Polycyclic Groups*, 127
Algebra and Applications 10,
DOI 10.1007/978-1-84882-941-1, © Springer-Verlag London Limited 2009